Fan Wu

Game Theoretic Approaches for Spectrum Redistribution

 Springer

Fan Wu
Department of Computer Science
and Engineering
Shanghai Jiao Tong University
Shanghai, People's Republic of China

ISSN 2191-8112   ISSN 2191-8120 (electronic)
ISBN 978-1-4939-0499-0   ISBN 978-1-4939-0500-3 (eBook)
DOI 10.1007/978-1-4939-0500-3
Springer New York Heidelberg Dordrecht London

Library of Congress Control Number: 2014931419

Printed on acid-free paper

Springer is part of Springer Science+Business Media (www.springer.com)

SpringerBriefs in Electrical and Computer Engineering

For further volumes:
http://www.springer.com/series/10059

Preface

This book presents a systematic study of game-theoretic approaches for dynamic spectrum redistribution. The objective of this book is to provide readers with a comprehensive understanding of fundamental design methodologies for efficient spectrum allocation with strategic participants. The book can be used as a graduate-level seminar textbook and as a reference for academic and industrial researchers and students working in this field.

In this book, Chap. 1 begins with an introduction of the problem of game-theoretic spectrum redistribution. A short tutorial of solution concepts in game theory is presented in Chap. 2. Chapter 3 covers state of the art in the literature. Chapter 4 provides two complementary game-theoretic approaches for channel allocation in a clique. In Chap. 5, the focus shifts to game-theoretic approaches in multi-hope wireless networks, where spectrum can be spatially reused. Finally, Chap. 6 summarizes the contents of this book and points out open problems.

The only prerequisite knowledge for this book is a basic understanding of game theory. For those who are not familiar with game theory, please refer to Chap. 2 for a short tutorial. A comprehensive introduction of game theory can be found in [46]. Background in wireless communication is recommended but not required, as the necessary underlying principles are covered in the game/auction models.

Many people's efforts result in the appearance of this book. I would like to thank Sheng Zhong, Chunming Qiao, Nitin Vaidya, Guihai Chen, Tingting Chen, Nikhil Singh, Tianrong Zhang, Qianyi Huang, Zhenzhe Zheng, Ruihao Zhu, and Yixin Tao for their insightful contributions to this book. I gratefully acknowledge the help from Xuemin (Sherman) Shen. Finally, thanks to Susan Lagerstrom-Fife, Courtney Clark, and Jennifer Malat for their valuable advice throughout the production of this book.

This work was supported in part by the State Key Development Program for Basic Research of China (grant no. 2014CB340303, 2012CB316201), in part by China NSF grant 61272443 and 61133006, and in part by Shanghai Science and Technology fund 12PJ1404900 and 12ZR1414900.

Shanghai, People's Republic of China   Fan Wu

Contents

Chapter 1
Introduction

The last two decades have witnessed a rapid development of wireless communication technology. Unfortunately, naturally limited radio spectrum is becoming a more and more serious bottleneck of ongoing growth of wireless applications and services. Most of the countries have specific departments to regulate spectrum usage, e.g., Federal Communications Commission (FCC) [19] in the US and Radio Administration Bureau (RAB) in China [48]. They statically allocate spectrum to wireless application service providers on a long term basis for large geographical regions. Such static management leads to low spectrum utilization in spatial and temporal dimensions. Large chunks of radio spectrum are left idle most of the time at a lot of places, while new wireless applications are starving for the radio spectrum. Therefore, an open and market-based framework is highly needed to dynamically redistribute the radio spectrum, and thus improve the utilization of the radio spectrum. Open markets, such as Spectrum Bridge [53], have already appeared to improve spectrum utilization by providing services for buying, selling, and leasing idle spectrum/channel.

In this book, we study the problem of spectrum/channel allocation from a game-theoretic perspective, in which the nodes in the wireless network are rational and always pursue their own objectives. Similar as most of the existing works, we first model the problem as a strategic game, and analyze the existence of Nash equilibrium (NE), which is a fundamental solution concept from game theory, when there is no exogenous factor to influence participants' behavior. However, NE does not provide an ideal solution to the problem of spectrum allocation in many cases. There are three reasons.

- NE is not a very strong solution concept. Specifically, when in a NE, a player of the game has incentives to keep its equilibrium strategy only under the assumption that all the other players are also keeping their equilibrium strategies. When this assumption is not valid, NE does not provide incentives for the game player.

F. Wu, *Game Theoretic Approaches for Spectrum Redistribution*, SpringerBriefs in Electrical and Computer Engineering, DOI 10.1007/978-1-4939-0500-3_1, © The Author(s) 2014

- More importantly, NE is usually not socially efficient, which means that the system performance is not optimized. Therefore, when the system converges to one of the NEs, it could be the case that some of the selfish players benefit at the cost of system performance degradation.
- Last but not least, the convergence to NE can be slow, or the system never converges at all. The system may shift among several states all the time.

The objective of this book is to provide a systematic study on the approaches that can guarantee the system to converge to an equilibrium state, in which the system performance is optimal or sub-optimal. Specifically, we use a very strong solution concept from game theory, called *Dominant Strategy Equilibrium*, to guarantee the system convergence. By its definition (please refer to Sect. 2), dominant strategy equilibrium ensures that a participant always has incentives to use the equilibrium strategy, regardless of others' behavior. Hence the solutions we provide are much stronger than any NE-based solution.

The rest of the book is organized as follows.

Chapter 2: A Short Tutorial on Game Theory

In this chapter, we recall the important notations and solution concepts from game theory. Game theory provides the basic theoretical building blocks for this book.

Chapter 3: State of Art on Channel Allocation

In this chapter, we briefly review the current state of art on spectrum allocation in wireless networks. We cover both cooperative channel allocation approaches, and channel allocation approaches dealing with selfish participants.

Chapter 4: Game-Theoretic Channel Allocation in Clique

In this chapter, we study the problem of game-theoretic channel allocation in clique, where all the nodes are in the same collision domain and have to compete with each other to get limited wireless media for communication.

In Sect. 4.1, we study the problem of allocating pre-partitioned channels in a non-cooperative wireless network, where devices are selfish. Existing work on this problem has considered the existence of and convergence to the NE, which is not a very strong solution concept and may not guarantee a good system performance. In contrast, we introduce a payment formula to ensure the existence of a Strongly Dominant Strategy Equilibrium (SDSE), which is a different solution concept that gives participants much stronger incentives. We show that, when the system converges to a SDSE, it also achieves global optimality in terms of system throughput.

In Sect. 4.2, we further consider the problem of allocating bandwidth adjustable channels, given that the width of IEEE 802.11-based communication channels can be changed adaptively in software, even by using commodity Wi-Fi hardware [8]. We first model the problem as a strategic game, and show the existence of NE, when there is no exogenous factor to influence players' behavior. We further propose a charging scheme to influence the players' behavior, by which the system is guaranteed to converge to a Dominant Strategy Equilibrium (DSE), a solution concept that gives participants much stronger incentives. We show that, when the system converges to a DSE, it also achieves global optimality, in terms of system-wide throughput without starvation.

Chapter 5: Game-Theoretic Channel Allocation in Multi-Hop Wireless Networks

In this chapter, we adopt the concept of auctions, which is the best-known market-based allocation mechanisms for redistributing resources, to perform highly efficient channel allocation in multi-hop wireless networks.

After presenting a basic auction model for channel allocation in Sect. 5.1, we first present a graph coloring-based auction mechanism for channel allocation, namely SMALL in Sect. 5.2. SMALL is a sealed-bid reserve auction mechanism, in which all bidders simultaneously submit sealed bids so that no bidder knows the bid of any of the other participants, and a channel may not be sold if the final bid is not high enough to satisfy the seller. We prove that SMALL is a strategy-proof auction mechanism.

In Sect. 5.3, we further present a combinatorial auction mechanism, namely SPECIAL, for the problem of channel allocation in multi-hop wireless networks. SPECIAL for the first time can take flexible bids for different numbers of contiguous channels, which efficiently capture the diverse needs of channel buyers. Our analysis again shows that SPECIAL is a strategy-proof channel auction mechanism.

In Sect. 5.4, we present a general model of combinatorial auction for hetero-geneous channel redistribution. The auction model is powerful enough to express channel spatial reusability and heterogeneity, as well as bid diversity. We then introduce the concept of *virtual channel* to capture the confliction of channel usage among different auction participants. By using virtual channels, we transform the problem of heterogeneous channel allocation to a classic multi-unit combinatorial auction. We present SMASHER, which is a combinatorial auction mechanism for heterogeneous channel redistribution, achieving both strategy-proofness and approximately efficient social welfare.

In Sect. 5.5, we study the problem of privacy preservation in strategy-proof spectrum auction mechanisms, which has been left open in the literature of game theory based spectrum redistribution for a long period of time. We present a practical auction mechanism, called PRIDE, to guarantee k-anonymous privacy preservation in a generic strategy-proof spectrum auction mechanism (e.g., [65, 82]). We also extend PRIDE to adapt to multi-channel bids, and let it still achieve both strategy-proofness and k-anonymity.

Chapter 6: Summary and Open Problems

Finally, in this chapter, we summarize the contributions presented in this book, and point out open problems.

Chapter 2
A Short Tutorial on Game Theory

Game theory [23, 25, 44–46] is a collection of analytical tools designed to study a system of self-interested decision-makers in conditions of strategic interaction. This section briefly reviews important game-theoretic solution concepts.

2.1 Strategic Games

2.1.1 Definition of Strategic Game

In the model of strategic game, there is a finite set of players $N = \{1, 2, \ldots, n\}$ and, for each player $i \in N$, a nonempty set Σ_i of all possible (mixed) strategies and a preference relation.

A strategy is a complete contingent plan, or decision rule, that defines the action a player will select in every distinguishable state of the game. A strategy can be either pure (deterministic), or mixed (stochastic). A mixed strategy $\sigma_i \in \Delta(\Sigma_i)$ defines a probability distribution over pure strategies. The set of strategy profiles is $\Sigma = \times_{i \in N} \Sigma_i$. Each player i chooses a strategy $s_i \in \Sigma_i$. As a notational convention, $s_{-i} = (s_1, \ldots, s_{i-1}, s_{i+1}, \ldots, s_n)$ represents the strategies of all players except player i. Note that $s = (s_i, s_{-i})$ is a strategy profile, in which player i takes strategy s_i and the other players take strategies s_{-i}.

A player i's preferences can be determined by a utility function $u_i(s)$ of the strategies of all the player. The utility function captures the essential concept of strategic interdependence. In other words, the utility, $u_i(s)$, of player i determines its preferences over its own strategy and the strategies of other players. Player i prefers strategy s_i to s_i' when the other players take s_{-i}, if $u_i(s_i, s_{-i}) > u_i(s_i', s_{-i})$.

F. Wu, *Game Theoretic Approaches for Spectrum Redistribution*, SpringerBriefs
in Electrical and Computer Engineering, DOI 10.1007/978-1-4939-0500-3_2,
© The Author(s) 2014

To summarize, the definition of the strategic game is the following.

Definition 2.1 (Strategic Game). A strategic game consists of

- a finite set $\mathbb{N}$ of players;
- for each player $i \in \mathbb{N}$ a nonempty set Σ_i of all possible (mixed) strategies;
- for each player $i \in \mathbb{N}$ a utility function $u_i(s)$, which determines the player's preference.

The strategic game models players' rationality in maximizing their individual expected utility. A player will always selects a strategy that maximized its expected utility, given its belief of the others players' strategies.

2.1.2   Nash Equilibrium

The most commonly used solution concept in game theory is *Nash Equilibrium* (NE) [46], which captures a steady state of the strategic game in which each player holds the correct expectation about the other players' behavior and acts rationally.

Definition 2.2 (Nash Equilibrium). A Nash Equilibrium of a strategic game is a profile $s^\star \in \Sigma$ of strategies with the property that for every player $i \in \mathbb{N}$ we have

$$u_i(s_i^\star, s_{-i}^\star) \geq u_i(s_i, s_{-i}^\star), \tag{2.1}$$

for all $s_i \in \Sigma_i$.

Intuitively, for any $s^\star$ to be a Nash Equilibrium, it must be that no player i has an action other than $s_i^\star$ yielding an outcome that is more beneficial to the player, given that every other player j chooses her equilibrium strategy $s_j^\star$. In other words, no player can get more benefit by unilaterally deviating from the Nash Equilibrium.

2.1.3   Dominant Strategy

Although the Nash Equilibrium gives a fundamental solution concept to game theory, it relies on knowing all the other players' strategies and beliefs on the other players, and also loses power in the games where multiple NEs exist. We now introduce a very strong solution concept called *dominant strategy* [46], in which every player has the utility-maximizing strategy for all strategies of the other players.

Definition 2.3 (Dominant Strategy). A dominant strategy of a player is one that maximizes its utility regardless of what strategies other players choose. Specifically, $s\star_i$ is player i's dominant strategy if, for any $s_i' \neq s\star_i$ and any s_{-i},

$$u_i(s\star_i, s_{-i}) \geq u_i(s_i', s_{-i}). \tag{2.2}$$

In other words, a strategy $s\star_i$ is a dominant strategy for a player if it maximizes her expected utility, no matter what strategies the other players take.

Dominant-strategy equilibrium is a very robust solution concept, because it makes no assumptions about the information available to players about each other, and does not require an player to believe that other players will behave rationally to select its own optimal strategy.

People further strengthen the requirement for dominant strategy equilibrium to have a stronger solution concept called *Strongly Dominant Strategy Equilibrium* (SDSE):

Definition 2.4 (Strongly Dominant Strategy Equilibrium). A Strongly Dominant Strategy Equilibrium of a strategic game is a profile $s^\star \in \Sigma$ of strategies with the property that for every player $i \in \mathbb{N}$,

$$\begin{cases} \forall s_{-i} \in \Sigma_{-i}, \forall s_i' \neq s_i^\star, \ u_i(s_i^\star, s_{-i}) \geq u_i(s_i', s_{-i}) \\ \exists s_{-i} \in \Sigma_{-i}, \forall s_i' \neq s_i^\star, \ u_i(s_i^\star, s_{-i}) > u_i(s_i', s_{-i}). \end{cases} \tag{2.3}$$

2.2 Mechanism Design

The mechanism design aims to implement an optimal system-wide solution to a decentralized optimization problem with self-interested players who have private information about their preferences for different outcomes. We call the private information of a player i as *type*, denoted by t_i.

A mechanism consists of the strategies available to the players and the method used to select the final outcome based on players' strategies. The mechanism design problem is to implement the rules of a game. Game theory is usually used to analyze the outcome of a mechanism.

A *direct-revelation* mechanism is a mechanism in which the only actions available to players are to make claims about their preferences to the mechanism. That is, the strategy of player i is reporting type $\hat{t}_i = s_i(t_i)$, based on its actual preferences t_i.

A direct-revelation mechanism is *incentive-compatible* (IC) if reporting truthful information is a dominant strategy for each player. Another important property of a mechanism is *individual-rationality* (IR)—each player can always achieve at least as much expected utility from participation as without participation. Finally, we say a direct-revelation mechanism is *strategy-proof* if it satisfies both IC and IR properties.

Definition 2.5 (Strategy-Proof Mechanism [40, 56]). A mechanism is strategy-proof when it satisfies both incentive-compatibility and individual-rationality.

Chapter 3
State of Art on Channel Allocation

In this chapter, we first review related works on channel allocation with the assumption of cooperation of participants, and then review the works involved with selfish participants.

3.1 Cooperative Channel Allocation

The problem of channel allocation was first studied in cellular networks. We refer to [36] for a comprehensive survey.

A number of works were focused on wireless LANs (WLANs). For example, in [41], Mishra et al. exploited weighted graph coloring to study the channel allocation for WLANs. In [42], Mishra et al. utilized client-driven mechanisms to study the joint problem of channel allocation and load balancing in centrally managed WLANs.

In wireless mesh networks (WMNs), the problems of channel allocation were also studied. For instance, Alicherry et al. [4], Raniwala et al. [50], and Kodialam and Naghshineh [37] combined channel allocation together with routing or scheduling, in order to achieve the maximum network throughput. Raman [49] focused on the channel allocation problem in rural mesh networks built with directional antennas.

The channel allocation problem is studied in other wireless networks as well, such as ad-hoc networks (e.g., [38, 57]), software defined radio networks (e.g., [30, 31]) and cognitive radio networks (e.g., [15, 77]).

3.2 Channel Allocation with Selfish Participants

All related works in Sect. 3.1 are based on the assumption that all nodes must be cooperative, which means that all nodes unconditionally comply with a central control or prescribed protocol. However, this assumption is not valid when the

F. Wu, *Game Theoretic Approaches for Spectrum Redistribution*, SpringerBriefs in Electrical and Computer Engineering, DOI 10.1007/978-1-4939-0500-3_3, © The Author(s) 2014

network consists of selfish nodes, whose goals are to maximize their benefits. Hence, the assignment of channels becomes a game.

Earlier, Felegyhazi et al. [21] studied Nash Equilibria in a static multi-radio multi-channel allocation game. Later, Wu et al. [67] designed a mechanism for the multi-radio multi-channel allocation game, converging to a much stronger equilibrium state, called strongly dominant strategy equilibrium (SDSE). These works considered the problem in a single collision domain. For multiple collision domain, Yang et al. [73] have formulated the channel allocation problem as a strategic game, called *ChAlloc*, and proposed a charging scheme to induce players to coverage to a Nash Equilibrium. In [68], Wu et al. have studied the problem of adaptive-width channel allocation. Another important work on channel allocation game is [29], where the authors proposed a graph coloring game model and discussed the price of anarchy under various topology conditions such as different channel numbers and bargaining strategies. Chen et al. proposed distributed spectrum sharing schemes to coverage Nash equilibrium in spectrum access game [9, 10]. In cognitive radio networks, Kasbekar et al. analyzed spectrum pricing game and computed Nash Equilibrium in different scenarios [34, 35].

Auctions are the most well-known market-based mechanisms to redistribute resources [32]. A number of strategy-proof auction-based spectrum allocation mechanisms (e.g., TRUST [82], VERITAS [84], SMALL [65] and SPECIAL [76]) have been proposed to solve the channel allocation problem. VERITAS and SMALL are single-sided auctions both supporting multiple channel requests. In contrast, TRUST elegantly extends double auction to consider both channel sellers and buyers' incentives. SPECIAL is the first work to consider the saturated throughput of contiguous channels. Recently, some researchers have considered the problem of heterogeneous channel redistribution, and have proposed a number of strategy-proof auction mechanisms [17, 22, 78]. Some other problems are also considered in the spectrum auction mechanism design, such as balancing social welfare and fairness [27], collusion-resistant in channel auction [83, 85], revenue generation for spectrum auction [3, 26, 33], and balancing social welfare and revenue [11]. Besides, there are also some works on online spectrum auction design (e.g., [14, 62, 69, 70, 72]).

Chapter 4
Game-Theoretic Channel Allocation in Clique

In recent years, a large number of channel allocation schemes for wireless networks (e.g., [4, 31, 36–38, 41, 42, 49, 50]) have been proposed. In general, they assumed that all the nodes are "well behaved" or "cooperative." However, this assumption may not be valid in general ad hoc networks [51]. In practice, a node can easily deviate from the protocol to seek for more benefit for itself. So it is crucial to study how to provide incentives for the selfish nodes to behave cooperatively. In a pioneer work, Felegyhazi et al. [20, 21] studied Nash Equilibria (NEs), which is a standard solution concept from game theory, in a non-cooperative multi-radio multi-channel allocation game. While their work is elegant and intriguing, NE does not provide an ideal solution to the problem of channel allocation, due to the three reasons listed in Chap. 1.

In this chapter, we study the problem of game-theoretic channel allocation in clique, where all the nodes are in the same collision domain and have to compete with each other to get limited wireless media for communication. Two complementary approaches are presented for pre-partitioned (Sect. 4.1) and adjustable (Sect. 4.2) channel allocation, respectively. Both of the two approaches can achieve globally optimal network throughput, and guarantee the system converge to a (strongly) dominant strategy equilibrium in a single step.

4.1 Optimal Allocation of Fixed Channels

In this section, we consider the problem of allocating previously partitioned channels, and present an approach that can guarantee the system to converge to a state in which the system throughput is optimized. Specifically, we use a very strong solution concept from game theory, called Strongly Dominant Strategy Equilibrium (SDSE), to guarantee the system convergence. By its definition (please refer to Sect. 2.1.3), SDSE ensures that, regardless of other nodes' behavior, a pair of

F. Wu, *Game Theoretic Approaches for Spectrum Redistribution*, SpringerBriefs in Electrical and Computer Engineering, DOI 10.1007/978-1-4939-0500-3_4, © The Author(s) 2014

communicating nodes always have incentives to use the equilibrium strategy. Hence the solution we provide is much stronger than any NE-based solution. The main technical tool we use in this work is a carefully designed payment scheme.

4.1.1   Strategic Game Model

We consider a static wireless network, where each node is equipped with one or multiple radio interfaces. We consider that pairs of nodes need to communicate with each other over a single hop, and each node participates in only one of the communication sessions. We assume that the communication pairs' packets are backlogged, which means that every communication pair has infinite packets to send. We denote the set of communication pairs by $N = \{1, 2, \ldots, n\}$, where each element i has an identification number. In this book, we assume the identification numbers are 1 through n.

To communicate, a pair of nodes need to tune one or multiple radios to their shared channel(s). We require that a transmission must be between two radios, where one acts as transmitter, and the other acts as receiver. So we only consider the case in which both nodes of a communication pair allocate the same number of radios to each of their shared channel(s). A pair of nodes can have parallel transmissions between them, if they both allocate more than one radios. Here, let each communication pair $i \in N$ have r_i radio pairs (i.e., both of the nodes have r_i radios). The radio pair distribution vector is denoted by $\mathbf{r} = (r_1, r_2, \ldots, r_n)$.

We assume there is a set of contiguous, orthogonal (non-interfering), and homogenous channels (e.g., 3 such channels in IEEE 802.11b/g and 24 in IEEE 802.11a), denoted by $\mathbb{C}$. The available channels are also numbered from 1 to $|\mathbb{C}|$. We further assume that $r_i \leq |\mathbb{C}|, \forall i \in N$. We consider MAC layer protocol used are CSMA/CA based protocols (e.g., IEEE 802.11 standards). We denote the aggregate throughput of a channel $c \in \mathbb{C}$ by $T(N_c)$, where N_c is the number of pairs of radio transmitter and receiver allocated to the channel c. Here, $T(N_c)$ is a non-increasing function of N_c. It can be either a constant independent of N_c or a decreasing function of N_c, corresponding to multiplexing scheme used. For instance, $T(N_c)$ is independent of N_c if TDMA based scheduling scheme is used; and $T(N_c)$ is a decreasing function when using CDMA or CSMA/CA based protocol (e.g., IEEE 802.11 standards). We assume that the aggregate throughput $T(N_c)$ of a channel c is shared evenly among the radios using the channel [6, 7]. So each radio pair gets throughput $T(N_c)/N_c$, when $N_c > 0$. Our analysis applies to any effective aggregate throughput function that satisfies above properties. Therefore, we do not specify a particular function here. One can get such a function through measurement in practice.

We model the problem of channel allocation as a strategic game, namely *channel allocation game*. In the channel allocation game, we consider the set N of communication pairs as players. In the rest of the book, we use player and communication pair interchangeably. The strategy of a player $i \in N$ is her channel allocation vector $s_i = (s_{i,1}, s_{i,2}, \ldots, s_{i,c}, \ldots, s_{i,|\mathbb{C}|})$, where $s_{i,c}$ is the number of radio pairs that player i assigns to channel c. The strategy profile s is a matrix composed of all the players' strategies: $s = (s_1, s_2, \ldots, s_n)^T$.

Given a strategy profile s, it is easy to see the total number of radio pairs used by a player i is $m_i(s) = \sum_{c \in \mathbb{C}} s_{i,c} \leq r_i$. Here, the inequality indicates that it is not necessary to use up all one's available radio pairs. Similarly, it is also easy to see the total number of radios assigned to a channel c is $N_c(s) = \sum_{i \in \mathbb{N}} s_{i,c}$. Hence, the throughput a player i gets from a channel c is

$$T_{i,c}(s) = \frac{s_{i,c}}{N_c} T(N_c), \tag{4.1}$$

and the total throughput a player i gets is

$$T_i(s) = \sum_{c \in \mathbb{C}} T_{i,c}(s). \tag{4.2}$$

Finally, the system throughput is:

$$T(s) = \sum_{c \in \mathbb{C}} T(N_c). \tag{4.3}$$

In reality, any practical solution to the channel allocation game should satisfy some additional requirements. First of all, there should not be any starvation. Second, we need *social efficiency*, which means that the system throughput should be maximized. We combine these two requirement to define the concept of global optimality of a solution[1]:

Definition 4.1 (Global Optimality). In a strategic game of channel allocation, suppose that s^* is a strategy profile or say a channel allocation. We say s^* is *globally optimal* if the following two requirements are met:

1. No starvation. $\forall i \in \mathbb{N}, T_i > 0$.
2. Social efficiency. $\forall s \neq s^*$, if s satisfies requirement (1), then $T(s) \leq T(s^*)$.

We note that the globally optimal channel allocation might not be unique. But all globally optimal channel allocations have the same overall throughput in the system.

4.1.2 Achieving Global Optimality

It is ideal to have a globally optimal channel allocation. However, achieving the globally optimal channel allocation is a highly challenging task. If we allow the

---

[1]Our definition of global optimality is thus slightly different from a traditional definition, which usually considers the optimization of a single metric (e.g., throughput).

players to choose the channels without giving them any influence, most likely the system would either not converge at all, or converge to a Nash equilibrium that is not globally optimal [20,21]. Therefore, we need to introduce a method to influence the strategies of the players. Here the method we use is to require players to make payments.

Just as in [5, 18, 59, 60, 79–81], we assume that there is some kind of virtual currency in the system. Each player has to pay some virtual money to the system administrator based on the outcome of the strategy profile. We regard this payment as the fee for using the channels. We note that the system administrator need not to be an online authority. It is just a server connected to the Internet. So the players can pay or receive credit from the system administrator when they have connections to the Internet.

Let's assume that we have a globally optimal strategy profile $s^\star$ (We will explain how to compute $s^\star$ in Sect. 4.1.3.). We define the payment of player i as follows:

$$p_i(s) = \alpha T_i(s) + \beta \left(D(s_i, s_i^\star) - \frac{1}{n-1} \sum_{j \in N \setminus \{i\}} D(s_j, s_j^\star) \right) - \epsilon, \qquad (4.4)$$

where $D(s_i, s_i^\star)$ is the Manhattan distance (also known as the L1-distance) between strategy profiles s_i and $s_i^\star$; $\alpha > 0$ and $\beta > 0$ are parameters used for converting throughput and the Manhattan distance into virtual currency values, respectively; $\epsilon > 0$ is a very small constant. (Suppose T_{min} is the minimal reasonable bandwidth, with which the basic networking operation can be completed. We require that $\epsilon < \alpha T_{min}$, so that $p_i > 0$ when all players follow the optimal strategy.) Intuitively, the payment is the charge for the player's overall throughput plus a penalty (bonus) for more (less) deviation from the globally optimal strategy than other players. We note that the total payments to the system administrator is:

$$P(s) = \sum_{i \in N} p_i(s) = \alpha \sum_{i \in N} T_i(s) - n\epsilon, \qquad (4.5)$$

which is the value of total throughput shared by the players minus $n\epsilon$. We further note that if all the channels are used,

$$P(s) = \alpha \sum_{i \in N} T_i(s) - n\epsilon = \alpha \sum_{c \in C} T(N_c) - n\epsilon, \qquad (4.6)$$

which is the value of system throughput minus $n\epsilon$.

We define the utility of player i as the value of throughput she obtains minus her payment to the system administrator:

$$u_i(s_i, s_{-i}) = \alpha T_i(s_i, s_{-i}) - p_i(s_i, s_{-i}). \qquad (4.7)$$

Since each player is selfish and rational, she always wants to maximize her utility.

Theorem 4.1. *It is a SDSE that each player i takes strategy $s_i^\star$.*

Proof. Combining Eqs. (4.4) and (4.7), we can get:

$$u_i(s_i, s_{-i}) = \epsilon - \beta \left(D(s_i, s_i^\star) - \frac{1}{n-1} \sum_{j \in N, j \neq i} D(s_j, s_j^\star) \right). \qquad (4.8)$$

Then the utility difference of taking strategy $s_i^\star$ and $s_i \neq s_i^\star$ is:

$$u_i^\star(s_i^\star, s_{-i}) - u_i(s_i, s_{-i})$$

$$= -\beta \left(D(s_i^\star, s_i^\star) - \frac{1}{n-1} \sum_{j \in N, j \neq i} D(s_j, s_j^\star) \right)$$

$$+ \beta \left(D(s_i, s_i^\star) - \frac{1}{n-1} \sum_{j \in N, j \neq i} D(s_j, s_j^\star) \right)$$

$$= \beta (D(s_i, s_i^\star) - D(s_i^\star, s_i^\star))$$

$$= \beta D(s_i, s_i^\star)$$

$$> 0. \qquad (4.9)$$

So strategy profile $s^\star$ is a SDSE. □

By Theorem 4.1, it is straightforward to see that, if $s^\star$ is a globally optimal channel allocation, then the SDSE achieved is also globally optimal.

4.1.3 Computing Globally Optimal Channel Allocation

To achieve the SDSE, each player must have an algorithm for computing the globally optimal allocation $s^\star$. In this subsection, we give a distributed algorithm that computes a globally optimal channel allocation $s^\star$, if there exists one.

The input of this algorithm is the set of channels $\mathbb{C}$, the set of players $\mathbb{N}$, and the radio distribution vector **r**. This algorithm requires perfect information of the system. Every player can obtain such information in ad hoc traffic indication message (ATIM) window of multi-channel MAC protocol (MMAC) [52] or channel switching function of multi-radio unification protocol (MUP) [1], by sending probe signals, which contain player ID and number of radio pairs, and listening to others' probe signals.

Algorithm 1 shows the pseudo-code of our algorithm. Intuitively, the algorithm considers three cases: (1) The number of players is less than that of channels. (2) The number of players is more than that of channels. (3) The number of players and that

---

Algorithm 1 Algorithm for computing globally optimal channel allocation

---

Input: Set of channels $\mathbb{C}$, set of players $\mathbb{N}$, radio distribution vector **r**.
Output: Globally optimal channel allocation s^*.
 1: Initialize all entries of s^* to 0.
 2: $i \leftarrow 1; c \leftarrow 1$.
 3: **while** $i \leq n$ and $c \leq |\mathbb{C}|$ **do**
 4:     $s^*_{i,c} \leftarrow 1; r_i \leftarrow r_i - 1$.
 5:     $i \leftarrow i + 1; c \leftarrow c + 1$.
 6: **end while**
 7: **if** $n < |\mathbb{C}|$ **then**
 8:     $i \leftarrow 1$.
 9:     **while** $c \leq |\mathbb{C}|$ and $i \leq n$ **do**
10:         **if** $r_i > 0$ **then**
11:             $s^*_{i,c} \leftarrow 1; r_i \leftarrow r_i - 1; c \leftarrow c + 1$.
12:         **else**
13:             $i \leftarrow i + 1$.
14:         **end if**
15:     **end while**
16: **else if** $n > |\mathbb{C}|$ **then**
17:     **while** $i \leq n$ **do**

18:         $c \leftarrow \underset{c \in \mathbb{C}}{argmin} \left(T\left(\sum_j s^*_{j,c}\right) - T\left(\sum_j s^*_{j,c} + 1\right) \right)$.

19:         $s^*_{i,c} \leftarrow 1$.
20:         $i \leftarrow i + 1$.
21:     **end while**
22: **end if**
23: **return** s^*.

---

of channels are equal. For all the cases, the algorithm first assigns each player with a single channel. Next, in case (1), the algorithm tries to assign each unoccupied channel with a player who still has unused radio pair, until all the channels are occupied or all the radios of players are used. In case (2), for each unassigned player i, the algorithm finds a channel c on which adding a radio pair will cause the least throughput degradation. Then it assigns player i with channel c. In case (3), we are done with channel allocation and the algorithm terminates.

It is not hard to see the correctness of Algorithm 1. In cases (1) and (3), since each channel is assigned with at most one radio pair, the allocation causes no throughput degradation. In case (2), since only one radio pair is used for each player and the allocation of each radio pair always causes the least throughput degradation, the overall throughput degradation is minimized. Putting these together, we can easily see that Algorithm 1 always computes a globally optimal channel allocation.

4.1.4 Evaluation Results

In this section, we evaluate our schemes using simulations. We assume that the available frequency band is divided into 12 orthogonal channels, which consist of

Table 4.1 Parameters used
to obtain numerical results

Packet payload	1,450 bytes
PHY&MAC header	50 bytes
ACK packet size	30 bytes
Minimum contention window	32
Number of backoff stages	5
Original channel bit rate	1 Mbps
Propagation delay	1 μs
Slot time	50 μs
SIFS	28 μs
DIFS	128 μs
ACK timeout	300 μs
$\alpha = \beta$	1

fixed-rate channels and varying-rate channels. In the evaluations, a basic CSMA/CA protocol with binary slotted exponential back-off is used for varying-rate channels. We use the same system parameters as those in work [6]. The parameters used to obtain numerical results are listed in Table 4.1.

Node Behavior: In our evaluations, we compare two kinds of node behavior:

- Following NE/SDSE: Following one's corresponding strategy in the NE/SDSE channel allocation strategy profile computed by our algorithms.
- Deviating NE/SDSE: Unilaterally deviating from one's corresponding strategy in the NE/DSE channel allocation strategy profile computed by our algorithms. For each of the radio pairs, the deviating player randomly chooses to use or not use the channel with equal probabilities.

Metrics: We evaluate two metrics:

- System throughput: System throughput is the sum of all the players' throughputs. This metric reflects the impacts of our design on the performance of the channel allocation game.
- Utility: Utility is the difference between the player's valuation on throughput and charge for using the channels. This metric reflects the impacts of a player's behavior on its own.

We have performed two sets of simulations. The first one is to compare the system throughput achieved by using our SDSE-based channel allocation scheme with that of random-based channel allocation scheme and NE-based channel allocation scheme [21]. In random-based scheme, players arbitrarily assign their radios to the channels. Here random-based scheme and NE-based scheme do not charge the players for using the channels. The second set of simulations is to show that if our scheme is used, deviating from the computed channel allocation cannot increase one's utility (see Eq. (4.7) for definition of utility).

In the first set of simulations, we consider three different channels deployments: (1) no varying-rate channel, (2) eight fixed-rate channels and four varying-rate channels, and (3) no fixed-rate channel. We vary the number of players from 2 to

20. The number of radio pairs each player has is uniformly distributed in $[1, 5]$. We repeat the simulation until the convergence level 10^{-6} is reached.

Figure 4.1 shows the result of the comparison on system throughputs between our SDSE-based scheme and the random-based scheme. Generally, the SDSE-based scheme reaches the maximum system throughput as long as there are only a small number of players. As Fig. 4.1 shows, in all three cases, the SDSE-based scheme reaches a system throughput of 12 Mbit/s with only 8 players. On the other hand, when there exist varying-rate channels, the system throughput of the random-based scheme will never reach 12 Mbit/s. Even without any varying-rate channel, the random-based scheme can get 12 Mbit/s only when there are at least 20 players in the system. Another advantage of the SDSE-based scheme is that it results in a much less system degradation than the random-based scheme, when there exist varying-rate channels. In Case (2), the SDSE-based scheme achieves a higher (0.68 Mbit/s more) system throughput than the random-based scheme in most cases; while in Case (3), the difference between system throughputs is as high as 1.76 Mbit/s or even more.

Fig. 4.1 System throughput achieved by using our SDSE-based scheme and the random-based scheme

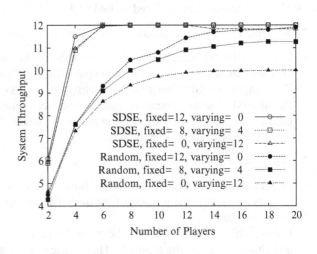

Figure 4.2 shows the comparison result between the SDSE-based scheme and NE-based scheme. Since there is no system degradation when no varying-rate channel exists, we only show the later two cases here. When the resource (channels) is abundant (less than or equal to 4 players, each with average of 3 radio pairs), the NE-based scheme achieves almost the same system throughput as the SDSE-based scheme. But when the resource is scarce, the greedy nature of the players in NE-based scheme will result in more severe contention for the channels as the number of players increases. Accordingly, the SDSE-based scheme performs much better than the NE-based scheme, when the resource is scarce. When there are 20 players, the system throughput of the SDSE-based scheme is 0.66 Mbps higher than that of the NE-based scheme for case (2), and 1.83 Mbps higher for case (3).

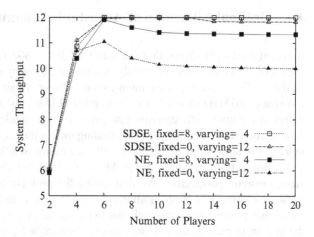

Fig. 4.2 System throughput achieved by using our SDSE-based scheme and the NE-based scheme

Our second set of simulations demonstrates the effect of some players deviating from our scheme. In this set of simulations, we assume that there are 20 players in the system, and 50 % of them deviate from our scheme by arbitrarily assigning their radios to the channels. The other setups are the same as the first set of simulations. The simulation is repeated 100 times. We keep track of a player and record her utility in the two cases: following our scheme or deviating from it.

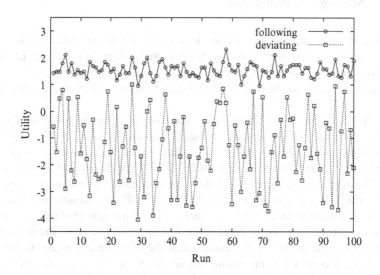

Fig. 4.3 Utility of following our scheme and deviating

Figure 4.3 illustrates the utility of the tracked player. It is shown that, when following our scheme, the player can always obtains non-negative utility. Furthermore, the utility obtained by following the scheme is always higher than by deviating from it. This will motivate each player to follow our scheme.

4.2 Optimal Allocation of Adjustable Channels

A recent study [8] shows that the width of IEEE 802.11-based communication channels can be changed adaptively in software, even by using commodity Wi-Fi hardware. For example, in communication between a pair of radio interfaces, two contiguous 20 MHz channels can be combined into a 40 MHz channel to provide higher throughput. Although much progress has been made in solving channel allocation problem in the literature, this finding makes the channel allocation problem more challenging—how to efficiently allocate radios to adjustable channels?

In this section, we study the problem of adjustable channel allocation from a game-theoretic perspective. We first model the problem as a strategic game, and show the existence of Nash equilibrium (NE). To overcome the weaknesses of NEs, we further present a charging scheme to influence the players' behavior, by which the system is guaranteed to converge to a Dominant Strategy Equilibrium (DSE). DSE ensures each player always has incentive to use the equilibrium strategy, regardless of the other nodes' strategies. We show that, when the system converges to the DSE, global optimality is also achieved. Therefore, the DSE-based outcome is better than the above NE-based outcome.

4.2.1 Strategic Game Model

Similar as the previous section, we consider a static wireless network, in which pairs of nodes need to communicate with each other over a single hop. In contrast to the previous scenario, in which the radios can only work on fixed width channels, we investigate the case that the radios can be tuned to operate on channels of arbitrary width.

As shown in paper [8], off-the-shelf wireless radio can be tuned to a wider channel combined by contiguous channels, and the bit-rate on the combined channel is proportional to its bandwidth. We consider MAC layer protocol used are CSMA/CA based protocols (e.g., IEEE 802.11 standards). Let B denote the bandwidth of a channel. We assume that we get a combined channel with bandwidth λB by combining λ contiguous channels. Let $T(W_c, N_c)$ denote the effective aggregate throughput on channel c, where c can be a single channel or a set of contiguous channels, W_c is the bandwidth of channel c, and N_c is the number of radio pairs competing for c. We assume that the effective aggregate throughput $T(W_c, N_c)$ is a convex non-increasing function of N_c for $N_c > 0$, when the width of channel W_c is fixed (refer to [6]); and $T(W_c, N_c)$ is a concave non-decreasing function of W_c for $W_c > 0$, when the number of competing radio pairs N_c is fixed (refer to [8]). Same as before, our analysis applies to any effective aggregate throughput function that satisfies above properties. Therefore, we do not specify a particular function here. One can get such a function through measurement in practice. If $c = \varnothing$ or $N_c = 0$, we define $T(W_c, N_c) = 0$. Figure 4.4 illustrates the trends of effective aggregate throughput $T(W_c, N_c)$ as a function of channel bandwidth W_c and the number of competing radio pairs N_c.

Since it is shown that the effective aggregate throughput on a channel c can be shared evenly among the radio transmitters using that channel [6,7], we assume that each radio pair on channel c gets a throughput $T(W_c, N_c)/N_c$, when $c \neq \emptyset$ and $N_c > 0$.

In the channel allocation game with adjustable channel, we still consider the set $\mathbb{N}$ of communication pairs as players. Each player $i \in \mathbb{N}$ has r_i radio pairs. The strategy of a player $i \in \mathbb{N}$ is its radio-channel allocation vector $s_i = (s_{i,1}, s_{i,2}, \ldots, s_{i,c}, \ldots, s_{i,|\mathbb{C}'|})$, where $\mathbb{C}'$ is a set of reorganized channels including single (original) channels and combined channels:

$$\bigcup_{c \in \mathbb{C}'} c = \mathbb{C}, \tag{4.10}$$

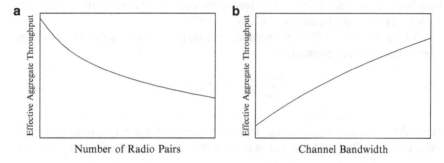

Fig. 4.4 Effective aggregate throughput $T(W_c, N_c)$ as a function of channel bandwidth W_c and the number of competing radio pairs N_c (**a**) Fixed channel bandwidth W_c (**b**) Fixed number of competing radio pairs N_c

and $s_{i,c}$ is the number of radio pairs that player i allocates on a channel $c \in \mathbb{C}'$. The reorganized channels in $\mathbb{C}'$ are also numbered from 1 to $|\mathbb{C}'|$. In this book, we do not consider the case in which the nodes cooperatively decide how to reorganize the channels. This cooperation leads to a coalitional game. Instead, we assume that $\mathbb{C}'$ is computed, independent of players strategies, by the system administrator, and then is broadcast to the players via the control channel. We leave the problem of channel allocation in cooperative wireless network as an open question. In contrast to the process of channel reorganization, channel allocation is done in a distributed manner.

Same as before, we use s to denote the matrix of all the players' strategies, and use s_{-i} to represent the strategy profile of all players except player i. Thus, $s = (s_i, s_{-i})$ is a complete strategy profile, in which player i takes strategy s_i and the other players take strategy profile s_{-i}.

Fig. 4.5 A toy example of channel allocation, where $\mathbb{N} = \{P_1, P_2, P_3\}$, $\mathbb{C}' = \{c_1', c_2'\}$ (here, $c_1' = \{c_1\}$, $c_2' = \{c_1, c_2\}$), and $r_1 = r_2 = r_3 = 2$

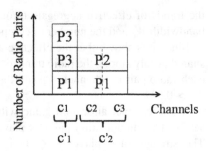

Figure 4.5 shows an example of channel allocation, with 3 players $\mathbb{N} = \{P_1, P_2, P_3\}$ and 3 original channels $\mathbb{C} = \{c_1, c_2, c_3\}$. Each player has 2 radio pairs. Channel 2 and channel 3 are combined into a wider channel, resulting in a new channel set $\mathbb{C}' = \{c_1', c_2'\}$, where $c_1' = \{c_1\}$ and $c_2' = \{c_1, c_2\}$. In Fig. 4.5, player 1 places its 2 radio pairs on channel c_1' and the combined channel c_2', respectively; player 2 only use 1 radio pair and tunes it on the combined channel c_2'; player 3 tune both of its radio pairs to c_1'. Hence, the strategies taken by the players are: $s_1 = (1, 1)$, $s_2 = (0, 1)$, and $s_3 = (2, 0)$.

Given a strategy profile s, the number of radio pairs used by a player i cannot be larger than its "radio constraint":

$$\sum_{c \in \mathbb{C}'} s_{i,c} \leq r_i, \forall i \in \mathbb{N}. \tag{4.11}$$

Here, the inequality indicates that it is not necessary for a player to use up all its radio pairs. Similarly, the number of radio pairs competing for a channel $c \in \mathbb{C}'$ is:

$$N_c(s) = \sum_{i \in \mathbb{N}} s_{i,c}. \tag{4.12}$$

Hence, the throughput a player i gets from a channel c is:

$$T_{i,c}(s) = \frac{s_{i,c}}{N_c(s)} \cdot T(W_c, N_c(s)), \tag{4.13}$$

and the total throughput a player i gets is:

$$T_i(s) = \sum_{c \in \mathbb{C}'} T_{i,c}(s). \tag{4.14}$$

Finally, the system throughput is:

$$T(s) = \sum_{i \in \mathbb{N}} T_i(s) = \sum_{c \in \mathbb{C}'} T(W_c, N_c(s)). \tag{4.15}$$

We assume that the players are rational and their objective is to maximize their own utilities. We denote the utility of a player $i \in \mathbb{N}$ by u_i:

$$u_i(s) = v_i(s) - p_i(s), \tag{4.16}$$

where $v_i(s)$ is player i's valuation on the outcome of the strategy profile s, and $p_i(s)$ is a charge for using the channels. For simplicity, we assume the valuation is proportional to the player's throughput:

$$v_i(s) = \alpha T_i(s), \tag{4.17}$$

where α is a positive constant coefficient. A well-designed charging formula can influence players' strategies and achieve good system-wide performance. In Sect. 4.2.2, we study the NE the system converges to, when there is no charging scheme to influence players' strategies (i.e., $p_i(s) = 0$). In Sect. 4.2.3, we present a charging scheme, and show that the system converges to a DSE, in which global optimality is also achieved.

4.2.2 Existence of Nash Equilibrium

In this section, we study the existence of Nash equilibrium, when there is no charging scheme to influence players' strategies. Hence, each player's objective becomes solely maximizing its throughput $u_i(s) = \alpha T_i(s)$.

We present an algorithm to compute a NE channel allocation strategy profile. The Nash equilibrium approach in this section is obtained using some modifications to the Nash equilibrium scheme in [21]. Every player must share the same algorithm for computing the channel allocation strategy profile s^* and the set of reorganized channels $\mathbb{C}^*$. We assume that every player can obtain the needed information by broadcasting beacons, containing one's identification number and number of radio pairs, and listening beacons from the other players, in the control channel.

4.2.2.1 Computing NE

Since contiguous channels can be combined into a wider channel in our model, we require that the computed set of reorganized channels should contain all the original channels.

Algorithm 2 shows the pseudo-code of our algorithm for computing a NE channel allocation strategy profile. In lines 2–14, we reorganize the channels by combining contiguous channels, if the total number of radio pairs is less than that of the channels, such that $\sum_{i \in \mathbb{N}} r_i \geq |\mathbb{C}^*|$. In this case, we create d combined channels, each containing $\lfloor |\mathbb{C}| / \sum_{i \in \mathbb{N}} r_i \rfloor + 1$ original channels, and $\sum_{i \in \mathbb{N}} r_i - d$ (combined) channels, each containing $\lfloor |\mathbb{C}| / \sum_{i \in \mathbb{N}} r_i \rfloor$ original channels. In lines 15–20, we

---

Algorithm 2 Algorithm for computing a NE channel allocation strategy profile

---

Input: A set of nodes $\mathbb{N}$, a set of channels $\mathbb{C}$, and a vector of radio pair distribution **r**.
Output: A channel allocation strategy profile s^* and a set of reorganized channels $\mathbb{C}^*$.
1: $\mathbb{C}^* \leftarrow \varnothing$; $s^* \leftarrow 0^{|\mathbb{N}|, |\mathbb{C}|}$.
2: **if** $\sum_{i \in \mathbb{N}} r_i < |\mathbb{C}|$ **then**
3: 　　$q \leftarrow \lfloor |\mathbb{C}| / \sum_{i \in \mathbb{N}} r_i \rfloor$; $d \leftarrow |\mathbb{C}| \mod \sum_{i \in \mathbb{N}} r_i$.
4: 　　$k \leftarrow 1$.
5: 　　**for** $j = 1$ to $\sum_{i \in \mathbb{N}} r_i$ **do**
6: 　　　　**if** $j \leq d$ **then**
7: 　　　　　　$\mathbb{C}^* \leftarrow \mathbb{C}^* \cup \{\{k, \ldots, k+q\}\}$; $k \leftarrow k+q+1$.
8: 　　　　**else**
9: 　　　　　　$\mathbb{C}^* \leftarrow \mathbb{C}^* \cup \{\{k, \ldots, k+q-1\}\}$; $k \leftarrow k+q$.
10: 　　　　**end if**
11: 　　**end for**
12: **else**
13: 　　$\mathbb{C}^* \leftarrow \mathbb{C}$.
14: **end if**
15: **for all** $i \in \mathbb{N}$ **do**
16: 　　**for** $j = 1$ to r_i **do**
17: 　　　　$c \leftarrow \underset{c \in \mathbb{C}^*}{argmin}(N_c(s^*))$.
18: 　　　　$s^*_{i,c} \leftarrow 1$.
19: 　　**end for**
20: **end for**
21: **return** s^* and $\mathbb{C}^*$.

---

evenly distribute all the radio pairs to the reorganized channels, and ensure that each player has at most 1 radio pair on any channel. Here, $argmin()$ function is a deterministic function. When there is a tie, the function return the channel with the smallest identification number. Finally, the algorithm returns a NE channel allocation strategy profile s^* and a set of reorganized channels $\mathbb{C}^*$.

4.2.2.2　Analysis

In this section, we prove that the channel allocation strategy profile s^* computed by Algorithm 2 is a NE on the set of reorganized channels $\mathbb{C}^*$. We distinguish two cases in our analysis.

1. $\sum_{i \in \mathbb{N}} r_i < |\mathbb{C}|$: The total number of radio pairs is less than the number of channels.
2. $\sum_{i \in \mathbb{N}} r_i \geq |\mathbb{C}|$: The total number of radio pairs is no less than the number of channels.

We first consider the case, where the total number of radio pairs is less than the number of channels, i.e., $\sum_{i \in \mathbb{N}} r_i < |\mathbb{C}|$. Considering the properties of the effective aggregate throughput function $T(W_c, N_c)$, to maximize the system-wide throughput, we should combine the original channels into $\sum_{i \in \mathbb{N}} r_i$ channels and allocate one radio pair to each of the reorganized channels. Due to the concavity of the effective aggregate throughput function $T(W_c, N_c)$ when N_c is fixed, the maximal system-wide throughput is achieved only when the original channels are evenly distributed into the combined channels:

$$\max\{W_c | c \in \mathbb{C}^*\} - \min\{W_c | c \in \mathbb{C}^*\} \le B. \tag{4.18}$$

We also show that it is a NE by allocating one radio pair to each of the reorganized channels in $\mathbb{C}^*$.

Lemma 4.1. *When* $\sum_{i \in \mathbb{N}} r_i < |\mathbb{C}|$, *channel allocation* s^* *on channel set* $\mathbb{C}^*$ *is a Nash equilibrium, if* $N_c(s^*) = 1, \forall c \in \mathbb{C}^*$.

Proof. Let s^* be a channel allocation on $\mathbb{C}^*$ satisfying $N_c(s^*) = 1, \forall c \in \mathbb{C}^*$. Since $\sum_{i \in \mathbb{N}} r_i = |\mathbb{C}^*|$ and $N_c(s^*) = 1, \forall c \in \mathbb{C}^*$, every player uses up all its radio pairs and allocate exactly one radio pair to each channel occupied.

Clearly, a player's utility decreases if it stops using any of its radio pairs. Next, we prove that a player cannot increase its utility by moving any of its radio pairs to any other channel. Let δ denote the utility difference of a player i, if it moves one radio pair from channel c to c', where $c, c' \in \mathbb{C}^*$:

$$\delta = \alpha T(W_c, 1) + \alpha s^*_{i,c'} T(W_{c'}, 1) - \frac{s^*_{i,c'} + 1}{2} \alpha T(W_{c'}, 2). \tag{4.19}$$

We consider the following two cases:

- $s^*_{i,c'} = 0$: Player i does not have a radio pair on channel c'.

$$\begin{aligned} \delta &= \alpha T(W_c, 1) - \frac{1}{2} \alpha T(W_{c'}, 2) \\ &\ge \alpha T(W_c, 1) - \alpha T(W_c, 2) \\ &\ge \alpha T(W_c, 1) - \alpha T(W_c, 1) \\ &= 0 \end{aligned} \tag{4.20}$$

- $s^*_{i,c'} = 1$: Player i has a radio pair on channel c'.

$$\begin{aligned} \delta &= \alpha T(W_c, 1) + \alpha T(W_{c'}, 1) - \alpha T(W_{c'}, 2) \\ &\ge \alpha T(W_c, 1) \\ &> 0 \end{aligned} \tag{4.21}$$

Therefore, the player cannot benefit by moving any of its radio pairs. Consequently, s^* is a NE on channel set $\mathbb{C}^*$. □

When $\sum_{i \in N} r_i < |\mathbb{C}|$, we have $\sum_{i \in N} r_i = |\mathbb{C}^*|$, and Algorithm 2 allocates exactly one radio pairs to each channel in $\mathbb{C}^*$. Therefore, s^* and $\mathbb{C}^*$ computed by Algorithm 2 satisfy Lemma 4.1, when $\sum_{i \in N} r_i < |\mathbb{C}|$.

We now consider the other case, where the total number of radio pairs is no less than that of the channels, i.e., $\sum_{i \in N} r_i \geq |\mathbb{C}|$. In this case, we do not combine any channel (i.e., $\mathbb{C}^* = \mathbb{C}$). A sufficient condition for NE was shown in [21]:

Lemma 4.2 ([21]). *When $\sum_{i \in N} r_i \geq |\mathbb{C}|$, a channel allocation s^* on channel set $\mathbb{C}^*$ is a Nash equilibrium, if it satisfies the following two conditions:*

1. $|N_c - N_{c'}| \leq 1, \forall c, c' \in \mathbb{C}^*$.
2. $s_{i,c}^* \leq 1, \forall i \in N, \forall c \in \mathbb{C}^*$.

Noting that line 17 and line 18 of Algorithm 2 indicate the two conditions in Lemma 4.2, respectively, s^* and $\mathbb{C}^*$ computed by Algorithm 2 also satisfy the sufficient condition shown in Lemma 4.2, when $\sum_{i \in N} r_i \geq |\mathbb{C}|$.

Combining Lemmas 4.1 and 4.2, we can get the following conclusion.

Theorem 4.2. *The channel allocation strategy profile s^* computed by Algorithm 2 is a NE on channel set $\mathbb{C}^*$.*

4.2.3   Achieving Global Optimality

As we have mentioned, NE has its weaknesses, and global optimality may not be achieved in NEs. Figure 4.6 shows an example, in which global optimality can be achieved by using another channel allocation, instead of the NE channel allocation computed in the previous section.

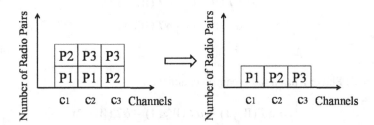

Fig. 4.6 An example of channel allocations, showing that system-wide throughput achieved by NE (*left*) can be improved by minimizing the number of players on each channel, using a better channel allocation (*right*). Here $N = \{P_1, P_2, P_3\}$, $\mathbb{C} = \{c_1, c_2, c_3\}$, and $r_1 = r_2 = r_3 = 2$

In this section, we overcome NE's weaknesses by introducing a carefully designed charging scheme, by which the system is strategy-proof and is guaranteed to achieve DSE. Hence, each player's objective is maximizing its utility, which is the difference between the player's valuation on its throughput and the charge for using the channels:

$$u_i(s) = \alpha T_i(s) - p_i(s). \tag{4.22}$$

We also show that the DSE achieves the global optimality in terms of system-wide throughput, under the condition of no starvation.

4.2.3.1 Charging Scheme

Our objective is to make the system converge to a globally optimal channel allocation. However, if we allow the players to choose the channels arbitrarily, most likely the system would either not converge at all, or converge to an state that is not globally optimal, as shown in the previous section. Therefore, we need to introduce a scheme in order to influence the strategies of the players. Here the scheme we use is to charge the players for using the channels.

As in the literature (e.g.,[18, 60, 79–81]), we assume that there is a virtual currency in the system. Each player has to pay some virtual money to the system administrator based on the outcome of the channel allocation game. We regard this charge as the fee for using the channels.

We now assume that we have a globally optimal channel allocation s^* on channel set $\mathbb{C}^*.$[2] (We will explain how to compute s^* in Sect. 4.2.4.)

Before introducing the charging scheme, we need to define a strategy subtraction operation "$\ominus$" between two strategies of a player i on channel $c \in \mathbb{C}^*$ and channel set $\mathbb{C}^*$:

$$s_{i,c} \ominus s'_{i,c} \triangleq (s_{i,c} - s'_{i,c} > 0) ? (s_{i,c} - s'_{i,c}) : 0, \tag{4.23}$$

$$s_i \ominus s'_i \triangleq ((s_{i,c} - s'_{i,c} > 0) ? (s_{i,c} - s'_{i,c}) : 0)_{c \in \mathbb{C}^*}. \tag{4.24}$$

Here, the strategy subtraction operation calculates the difference between each pair of corresponding elements in the two strategies. If the difference is not positive, then the operation lets the result element be 0. Intuitively, the strategy subtraction operation calculates the difference between two channel allocation strategies, without resulting in any negative number of radio pairs allocated onto a channel.

---

[2]Given a globally optimal channel allocation, a straightforward way to enforce the channel allocation is to punish any deviating player with an infinite charge. However, such an approach may hurt the players' willingness to participate the channel allocation game. For example, a player may not deliberately deviate, but may happen to somewhat deviate from the globally optimal channel allocation, because her information is incomplete or inconsistent with others when computing the channel allocation. Therefore, we present an alternative charging scheme in this book.

Let $\hat{s}_i(\hat{s}_{i,c})$ denote the difference between player i's strategy $s_i(s_{i,c})$ and its strategy in the globally optimal channel allocation $s^*(s^*_{i,c}$, on channel c), respectively:

$$\hat{s}_i = s_i \ominus s^*_i, \tag{4.25}$$

$$\hat{s}_{i,c} = s_{i,c} \ominus s^*_{i,c}. \tag{4.26}$$

We now define the charge of player i as follows:

$$p_i(s) = \alpha \left(T(s_i \ominus \hat{s}_i, s_{-i}) - T(s) + \bar{T}_i(s) + \hat{T}_i(s) \right), \tag{4.27}$$

where

$$\bar{T}_i(s) = \sum_{c \in \mathbb{C}^* \wedge N_c(s) = \hat{s}_{i,c}} T(W_c, \hat{s}_{i,c}), \tag{4.28}$$

$$\hat{T}_i(s) = \sum_{c \in \mathbb{C}^*} \left(\frac{\hat{s}_{i,c}}{N_c(s)} \cdot T(W_c, N_c(s)) \right). \tag{4.29}$$

Intuitively, there are three parts within bigger parentheses in the charging formula. The first part $T(s_i \ominus \hat{s}_i, s_{-i}) - T(s)$ calculates the degradation of system-wide throughput, if the player i deviate from s^*_i by allocating extra radio pairs specified in $\hat{s}_i$ to the channels. We note that this part can be negative if the player i increases system-wide throughput by using strategy s_i instead of s^*_i, when the other players take strategy profile s_{-i}. This happens when the player i luckily allocates an extra radio pair to a *free channel* (i.e., a channel not occupied by any other radio pairs specified in $(s_i \ominus \hat{s}_i, s_{-i})$). The second part $\bar{T}_i(s)$ represents the throughput got by the player i using extra radio pairs specified in $\hat{s}_i$ on the free channel(s). Putting the first two parts together, we get the throughput degradation on channels that are not free. Since the second part cancels the possible system-wide throughput increment achieved by player i by allocating radios to the free channels, we note that the sum of first two parts is always non-negative (i.e., $T(s_i \ominus \hat{s}_i, s_{-i}) - T(s) + \bar{T}_i(s) \geq 0$). The third part $\hat{T}_i(s)$ is the total throughput got by the extra radio pairs specified in $\hat{s}_i$. We also note that, if $s_i = s^*_i$, then $p_i = 0$; otherwise, $p_i \geq 0$.

Before proving that the channel allocation strategy profile s^* is a DSE on channel set $\mathbb{C}^*$, we first show the following lemmas.

Lemma 4.3. *If a channel allocation strategy profile s^* is globally optimal on channel set $\mathbb{C}^*$, then $s^*_{i,c} \in \{0, 1\}, \forall i \in \mathbb{N}, \forall c \in \mathbb{C}^*$.*

Proof. We prove this lemma by contradiction. Suppose a channel allocation strategy profile s^* is globally optimal on channel set $\mathbb{C}^*$, and there exist $i \in \mathbb{N}$ and $c \in \mathbb{C}^*$, such that $s^*_{i,c} > 1$. Then we can simply increase the effective aggregate throughput on channel c, by removing one radio pair of the player i from channel c. Thus, the system-wide throughput can be improved, without starving the player i. Here comes the contradiction.   □

Lemma 4.4. *In the charge formula, we have*

$$T(s_i \ominus \hat{s}_i, s_{-i}) - T(s) + \bar{T}_i(s) \geq 0. \tag{4.30}$$

Proof. We consider two cases:

- $\forall c \in \mathbb{C}^\star, s_{i,c} \leq s_{i,c}^\star$: In this case,

$$s_i \ominus \hat{s}_i = s_i, \tag{4.31}$$

and therefore the difference of the first two terms is zero

$$T(s_i \ominus \hat{s}_i, s_{-i}) - T(s) = 0. \tag{4.32}$$

Since $\hat{s}_{i,c} = s_{i,c} \ominus s_{i,c}^\star$, we have $\bar{T}_i(s) = 0$. Therefore,

$$T(s_i \ominus \hat{s}_i, s_{-i}) - T(s) + \bar{T}_i(s) = 0. \tag{4.33}$$

- $\exists c \in \mathbb{C}^\star, s_{i,c} > s_{i,c}^\star$: In this case, for $c \in \mathbb{C}^\star$, s.t. $s_{i,c} > s_{i,c}^\star$, we have $s_{i,c} \ominus \hat{s}_{i,c} = s_{i,c}^\star$.

$$
\begin{aligned}
&T(s_i \ominus \hat{s}_i, s_{-i}) - T(s) + \bar{T}_i(s) \\
&= \sum_{c \in \mathbb{C}^\star \wedge s_{i,c} > s_{i,c}^\star} \left(T(W_c, N_c(s_i \ominus \hat{s}_i, s_{-i})) - T(W_c, N_c(s)) \right) \\
&\quad + \sum_{c \in \mathbb{C}^\star \wedge N_c(s) = \hat{s}_{i,c}} T(W_c, \hat{s}_{i,c})
\end{aligned}
\tag{4.34}
$$

The system throughput can be lower than that of $T(s)$ if s_i is replaced by $s_i^\star$ (i.e., $T(s_i^\star, s_{-i}) - T(s) < 0$), when there was an empty channel (zero radio initially) on which a radio was placed in s_i. In such case, $\bar{T}_i(s) > 0$. After canceling the throughput $\bar{T}_i(s)$ got by player i on empty channel(s), we get the degradation of system throughput on non-empty channels as follows:

$$
\begin{aligned}
(4.34) \\
&= \sum_{c \in \mathbb{C}^\star \wedge s_{i,c} > s_{i,c}^\star \wedge N_c(s) \neq \hat{s}_{i,c}} \left(T(W_c, N_c(s_i \ominus \hat{s}_i, s_{-i})) - T(W_c, N_c(s)) \right) \\
&\geq 0
\end{aligned}
\tag{4.35}
$$

Putting the two cases together, we get $T(s_i \ominus \hat{s}_i, s_{-i}) - T(s) + \bar{T}_i(s) \geq 0$.   □

We now can get the following theorem.

Theorem 4.3. *If our charging scheme is used, then it is a DSE when each player i takes channel allocation strategy s_i^* on channel set $\mathbb{C}^*$.*

Proof. By combining Eqs. (4.22) and (4.27), we get an extended form of the utility function:

$$u_i(s) = \alpha\left(T_i(s) - T(s_i \ominus \hat{s}_i, s_{-i}) + T(s) - \bar{T}_i(s) - \hat{T}_i(s)\right). \qquad (4.36)$$

Then a player i's utility difference of taking strategy s_i^* and $s_i \neq s_i^*$, for any strategy profile s_{-i} of the other players, is shown as follows:

$$u_i^*(s_i^*, s_{-i}) - u_i(s_i, s_{-i})$$

$$= \alpha T_i(s_i^*, s_{-i}) - \alpha\left(T_i(s_i, s_{-i}) - T(s_i \ominus \hat{s}_i, s_{-i}) + T(s_i, s_{-i})\right.$$

$$\left. - \bar{T}_i(s_i, s_{-i}) - \hat{T}_i(s_i, s_{-i})\right)$$

$$= \alpha\left(T_i(s_i^*, s_{-i}) - T_i(s_i, s_{-i}) + \hat{T}_i(s_i, s_{-i})\right)$$

$$+ \alpha\left(T(s_i \ominus \hat{s}_i, s_{-i}) - T(s_i, s_{-i}) + \bar{T}_i(s_i, s_{-i})\right)$$

$$\geq \alpha\left(T_i(s_i^*, s_{-i}) - T_i(s_i, s_{-i}) + \hat{T}_i(s_i, s_{-i})\right)$$

$$= \alpha \sum_{c \in \mathbb{C}^*}\left(\frac{s_{i,c}^*}{N_c(s_i^*, s_{-i})}T(W_c, N_c(s_i^*, s_{-i})) - \frac{s_{i,c} - \hat{s}_{i,c}}{N_c(s_i, s_{-i})}T(W_c, N_c(s_i, s_{-i}))\right)$$

$$\tag{4.37}$$

Let

$$Z_{i,c} = \frac{s_{i,c}^*}{N_c(s_i^*, s_{-i})}T(W_c, N_c(s_i^*, s_{-i})) - \frac{s_{i,c} - \hat{s}_{i,c}}{N_c(s_i, s_{-i})}T(W_c, N_c(s_i, s_{-i})). \quad (4.38)$$

Then

$$u_i^*(s_i^*, s_{-i}) - u_i(s_i, s_{-i}) = \alpha \sum_{c \in \mathbb{C}^*} Z_{i,c}. \qquad (4.39)$$

From Lemma 4.3, we get that $s_{i,c}^* \in \{0, 1\}$. We now consider two cases:

- $s_{i,c}^* = 0$: In this case,

$$s_{i,c} - \hat{s}_{i,c} = s_{i,c} - (s_{i,c} \ominus s_{i,c}^*)$$

$$= s_{i,c} - s_{i,c}$$

$$= 0. \qquad (4.40)$$

Therefore, $Z_{i,c} \geq 0$.

- $s_{i,c}^{\star} = 1$: In this case, $0 \leq s_{i,c} - \hat{s}_{i,c} \leq 1$. We further distinguish two cases:

  - If $s_{i,c} - \hat{s}_{i,c} = 0$, clearly, $Z_{i,c} \geq 0$.
  - If $s_{i,c} - \hat{s}_{i,c} = 1$, then $s_{i,c} \geq 1$. Consequently, we have

$$N_c(s_i, s_{-i}) \geq N_c(s_i^{\star}, s_{-i}), \tag{4.41}$$

and

$$T(W_c, N_c(s_i, s_{-i})) \leq T(W_c, N_c(s_i^{\star}, s_{-i})). \tag{4.42}$$

Therefore,

$$Z_{i,c} \geq \frac{s_{i,c}^{\star} - (s_{i,c} - \hat{s}_{i,c})}{N_c(s_i^{\star}, s_{-i})} T(W_c, N_c(s_i^{\star}, s_{-i})) = 0. \tag{4.43}$$

All in all, $u_i^{\star}(s_i^{\star}, s_{-i}) - u_i(s_i, s_{-i}) \geq 0$. So the channel allocation strategy profile $s^{\star}$ is a DSE on channel set $\mathbb{C}^{\star}$. $\square$

By Theorem 4.3, it is straightforward to conclude that, if $s^{\star}$ is a globally optimal channel allocation strategy profile, then the DSE achieved is also globally optimal.

4.2.4 Computing Globally Optimal Channel Allocation

To implement the DSE, every player must share the same algorithm for computing the globally optimal channel allocation strategy profile $s^{\star}$ and the set of reorganized channels $\mathbb{C}^{\star}$. In this section, we present an algorithm that requires perfect information of the network. We assume every player can obtain such information by broadcasting beacons, containing one's identification number and number of radio pairs, and listening beacons from the other players, in the control channel.

Algorithm 3 shows the pseudo-code of our algorithm for computing a globally optimal channel allocation strategy profile. The first part of the algorithm (lines 2–14) is identical to that of Algorithm 2, which reorganize the channels according to the total number of radio pairs, such that $\sum_{i \in \mathbb{N}} r_i \geq |\mathbb{C}^{\star}|$. Next, the algorithm evenly distribute the players onto the reorganized channels (lines 15–18). If the number of players is less than that of the channels, the algorithm keeps allocating each unoccupied channel with a player who still has unused radio pair, until all the channels are occupied (lines 19–26). For fairness, the algorithm tries to make every player has almost the same number of radio pairs allocated, unless it does not has sufficient number of radio pairs. Finally, the algorithm returns a globally optimal channel allocation strategy profile $s^{\star}$ and a set of reorganized channels $\mathbb{C}^{\star}$. The computational complexity of Algorithm 2 is $O(n|\mathbb{C}|)$.

It is straightforward to see the correctness of Algorithm 3, because the algorithm always fully utilizes all the channels and causes minimal system-wide throughput degradation, while ensuring that no player get starved.

Algorithm 3 Algorithm for computing a globally optimal channel allocation strategy profile

---

Input: A set of nodes $\mathbb{N}$, a set of channels $\mathbb{C}$, and a vector of radio pair distribution $\mathbf{r}$.
Output: A channel allocation strategy profile $s^\star$ and a set of reorganized channels $\mathbb{C}^\star$.
 1: $\mathbb{C}^\star \leftarrow \varnothing$; Initialize $s^\star$.
 2: **if** $\sum_{i \in \mathbb{N}} r_i < |\mathbb{C}|$ **then**
 3:     $q \leftarrow \lfloor |\mathbb{C}| / \sum_{i \in \mathbb{N}} r_i \rfloor; d \leftarrow |\mathbb{C}| \mod \sum_{i \in \mathbb{N}} r_i$.
 4:     $k \leftarrow 1$.
 5:     **for** $j = 1$ to $\sum_{i \in \mathbb{N}} r_i$ **do**
 6:         **if** $j \le d$ **then**
 7:             $\mathbb{C}^\star \leftarrow \mathbb{C}^\star \cup \{\{k, \ldots, k+q\}\}; k \leftarrow k+q+1$.
 8:         **else**
 9:             $\mathbb{C}^\star \leftarrow \mathbb{C}^\star \cup \{\{k, \ldots, k+q-1\}\}; k \leftarrow k+q$.
10:         **end if**
11:     **end for**
12: **else**
13:     $\mathbb{C}^\star \leftarrow \mathbb{C}$.
14: **end if**
15: **for all** $i \in \mathbb{N}$ **do**
16:     $c \leftarrow \underset{c \in \mathbb{C}^\star}{argmin}(N_c(s^\star))$.
17:     $s^\star_{i,c} \leftarrow 1$.
18: **end for**
19: $i \leftarrow 0$.
20: **while** $\exists c \in \mathbb{C}^\star, N_c(s^\star) = 0$ **do**
21:     **repeat**
22:         $i \leftarrow (i \mod |\mathbb{N}|) + 1$.
23:     **until** $\sum_{c \in \mathbb{C}^\star} s^\star_{i,c} < r_i$
24:     $c \leftarrow \underset{c \in \mathbb{C}^\star}{argmin}(N_c(s^\star))$.
25:     $s^\star_{i,c} \leftarrow 1$.
26: **end while**
27: **return** $s^\star$ and $\mathbb{C}^\star$.

Theorem 4.4. *The channel allocation strategy profile $s^\star$ and the set of reorganized channels $\mathbb{C}^\star$ computed by Algorithm 3 is globally optimal.*

4.2.5 Evaluation Results

We evaluate the channel allocation outcome achieved in the DSE, the SDSE [67], and the NE. The objective of our simulations is twofold. One is to verify that our charging scheme indeed overcomes the weaknesses of NE, and ensures that the system achieves DSE. The other one is to measure the influence of our scheme on the system performance, in terms of system-wide throughput.

In the evaluations, a basic CSMA/CA protocol with binary slotted exponential back-off is used for MAC layer protocol. The values of the parameters used to obtain numerical results are same as those in the previous section, and are listed in Table 4.1. Our charging scheme is applied only when evaluating the DSE, while not for evaluating the NE. When the SDSE-based scheme is evaluated, there is no channel combination.

Our first set of simulations is to evaluate system performance, in terms of system-wide throughput, when our charging scheme is applied or not. As shown in previous sections, with our charging scheme, the system converges to a DSE; otherwise, the system converges to a NE. Here, we use DSE-based and NE-based outcomes to represent the results of the two cases, respectively. In addition, we also compare the outcomes of our scheme with that of the SDSE-based scheme. In the evaluation, we repeat each simulation until the convergence level 10^{-6} is reached.

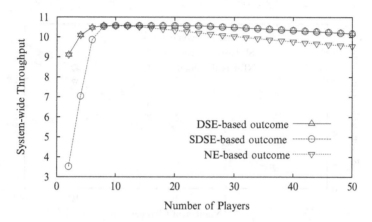

Fig. 4.7 System-wide throughputs achieved by the DSE-based, the SDSE-based, and the NE-based outcomes as a function of number of players, when the number of channels is 12 and the average number of radio pairs per player is 2

Figure 4.7 shows the system-wide throughputs achieved by the DSE-based, the SDSE-based, and the NE-based outcomes as a function of number of players, where the number of channels is 12 and the number of radio pairs on each player is randomly selected in $[1, 3]$. The figure shows that the DSE-based outcome always achieve the highest system-wide throughput. Specifically, when the number of players is no more than 12 (resource is relatively abundant to competitors), the NE-based and the DSE-based outcomes achieve almost the same system-wide throughput. However, the selfish nature of the players makes the performance of the NE-based outcome more and more worse than that of the DSE-based outcome, with the growth of the number of players. Specifically, when there are 50 players, the system-wide throughput of the DSE-based outcome is 0.63 Mbps (or 6.59 %) higher than that of the NE-based outcome. We can also observe that although the SDSE-based outcome has almost the same system-wide throughput with that of the DSE-based outcome, the SDSE-based outcome performs very badly when the number of the players is small (i.e., less than 8). This is because the SDSE-based scheme cannot efficiently utilize all the channels when the number of radio pairs is less than that of the channels.

Figures 4.8 and 4.9 show the system-wide throughputs achieved by the DSE-based, the SDSE-based, and the NE-based outcomes as a function of number of channels, in two different cases. Figure 4.8 presents the results when the spectrum resource is relatively scarce (i.e., the number of radio pairs is much larger than that of the channels), while Fig. 4.9 presents the results when the spectrum resource is relatively abundant (i.e., the number of radio pairs may be less than that of the channels). Again, these results show that the DSE-based outcome always performs at least as good as the other outcomes. In particular, when the spectrum resource is scarce as shown in Fig. 4.8, the DSE-based and the SDSE-based outcomes achieve almost the same system-wide throughputs, and have superior performance to that

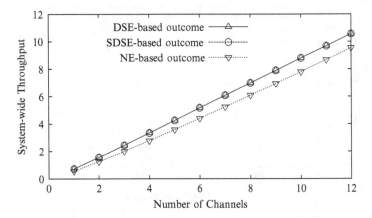

Fig. 4.8 System-wide throughputs achieved by the DSE-based, the SDSE-based, and the NE-based outcomes as a function of number of channels, when the number of players is 20 and the average number of radio pairs per player is 5

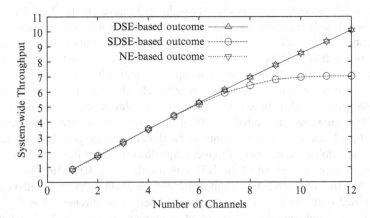

Fig. 4.9 System-wide throughputs achieved by the DSE-based, the SDSE-based, and the NE-based outcomes as a function of number of channels, when the number of players is 4 and the average number of radio pairs per player is 2

of the NE-based outcome. When the spectrum resource is abundant, the DSE-based
and the NE-based outcomes have better performance than the SDSE-based outcome.

Figure 4.10 show the system-wide throughputs achieved by the DSE-based, the
SDSE-based, and the NE-based outcomes as a function of average number of radio
pairs per player, when there are 10 players and 12 channels. These results once
again show that the DSE-based outcome always performs the best in the evaluation.
The SDSE-based outcome only performs well when the average number of radio
pairs per player is no less than 3, while the NE-based outcome only achieves good
system-wide throughput when the average is 1.

The above results show that the DSE-based outcome is more desirable than the
other outcome. In other words, the usage of our scheme is important in improving
system-wide throughput.

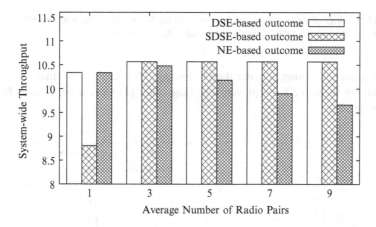

Fig. 4.10 System-wide throughputs achieved by the DSE-based, the SDSE-based, and the NE-
based outcomes for different average numbers of radio pairs per player, when the number of players
and channels are is 10 and 12, respectively

In our second set of simulations, we evaluate the impacts of a player's behavior
on its own with or without our charging scheme, and compare the strength of the
incentives provided by NE and DSE. In this set of simulations, we assume there
are 20 players and 12 channels in the system. The number of radio pairs on each
player is uniformly selected in [1, 3]. Each simulation is repeated 1,000 times. Due
to the limitation of space, we only show the results of the first 50 runs. In each run,
we record the players' utilities obtained by following NE/DSE and deviating from
NE/DSE, while the other players' strategy profile does not change.

Figure 4.11 shows the utilities of a randomly selected player (player 6) obtained
by following the NE and deviating from the NE, when the other players deviate
from the NE channel allocation strategy profile. We note that the results for the
other players are similar to that of player 6. The figure shows that, in some runs, the
utility of following the NE is no less than that of deviating from the NE; while in

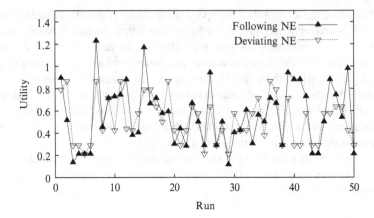

Fig. 4.11 Utilities of player 6 obtained by following the NE vs. deviating from the NE, when the other players deviate from the NE channel allocation strategy profile

the other runs, deviating is better than following. This result shows that NE cannot provide any incentive for a player, when the other players do not follow the NE strategy profile.

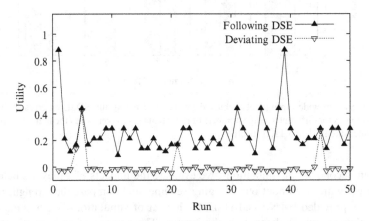

Fig. 4.12 Utilities of player 6 obtained by following the DSE vs. deviating from the DSE, when the other players deviate from the DSE channel allocation strategy profile

Figure 4.12 shows the utilities of a the same player (player 6) obtained by following the DSE and deviating from the DSE, when the other players deviate from the DSE channel allocation strategy profile. The figure shows that the utility of a player while following the DSE is always no lower than that of deviating. Furthermore, the utility of following is always positive, while that of deviating is negative most of the time. Therefore, when our charging scheme is used, the incentives for using the DSE strategy is always guaranteed, no matter what the other players' strategy profile is. This result verifies that players cannot benefit by deviating from the DSE channel allocation strategy profile, when our charging scheme is used. So that our charging scheme indeed guarantee the convergence to the DSE.

We observe that the peak utility of a player in Fig. 4.11 is higher than that in Fig. 4.12 (i.e., 1.23 in Fig. 4.11 and 0.88 in Fig. 4.12). This is because a player always wants to use up all of its radio pairs, when our charging scheme is not applied and there exist other competing player. However, this higher peak utility is achieved as a cost of system-wide throughput degradation, which will be shown in the next set of evaluations.

Chapter 5
Game-Theoretic Channel Allocation in Multi-Hop Wireless Networks

The previous globally optimal schemes cannot be used in multi-hope wireless networks, in which spectrum/channels can be spatially reused, due to computation intractability [65,66]. An alternative and effective way to redistribute radio spectrum is to use auction, which is a process of buying and selling goods by offering them up for bid, taking bids, and then selling the item(s) to the highest bidder(s). Since 1994, the Federal Communications Commission (FCC) has conducted auctions of licenses for radio spectrum [19]. While FCC auctions target only large wireless applications, we consider small wireless application buyers, such as community wireless networks and home wireless networks. These small buyers can search for and reuse idle chunks of radio spectrum.

However, designing a feasible channel auction mechanism has its own challenges. The first challenge, which is not only limited to channel auctions but applies to auctions in general, is strategy-proofness meaning that by reporting true valuation of the good as the bid, each buyer can maximize her payoff. Since the buyers always want to maximize their own payoff, they may manipulate the auction to seek for more benefit if the auction mechanism is not properly designed. Such misbehavior may hurt the benefit of truthful buyers, and thus discourage truthful buyers from participating in the auction. The second challenge is the efficiency of the channel allocation. Different from conventional goods, wireless channels have a property of spatial reusability, which means that wireless users that are well geographically separated can use the same channel simultaneously. Even if there is a powerful central authority, computing the optimal channel allocation is NP-complete in a multi-hop wireless network [13,75].

In this chapter, we adopt the concept of auctions to perform highly efficient channel allocation in multi-hop wireless networks, in which the radio spectrum can be spatially reused. Four strategy-proof channel auction mechanisms have been presented, including SMALL (Sect. 5.2), SPECIAL (Sect. 5.3), SMASHER (Sect. 5.4), and PRIDE (Sect. 5.5). SMALL is a graph coloring-based sealed-bid reserve auction mechanism, which accepts uniform bids for a single or multiple channels. SPECIAL is a combinatorial channel auction mechanism accepting flexible bids for different

F. Wu, *Game Theoretic Approaches for Spectrum Redistribution*, SpringerBriefs in Electrical and Computer Engineering, DOI 10.1007/978-1-4939-0500-3_5,

numbers of contiguous channels. SMASHER is a combinatorial auction mechanism for heterogeneous channel redistribution, achieving approximately efficient social welfare. Finally, PRIDE is a privacy preserving and strategy-proof channel auction mechanism.

5.1 Auction Model

We consider a static scenario in which there is a large wireless service provider, called "seller", who possesses a number of orthogonal spectrum channels and wants to lease out regionally unused channels; and there is a set of static nodes, called "buyers", such as WiFi access points, who want to lease channels in order to provide services to their users. A channel can be leased to multiple buyers, if these buyers can transmit simultaneously and receive signals with an adequate signal to interference and noise ratio (SINR).

We model this problem as a sealed-bid auction, in which all buyers simultaneously submit sealed bids so that no buyer knows the bid of any other participant. The objective of the auction is to efficiently allocate the channels to the buyers based on their bids, without violating interference conditions between the buyers.

We assume that the seller is trustworthy, and has a set $\mathbb{C} = \{c_1, c_2, \ldots, c_m\}$ of orthogonal and homogenous channels to lease. Each channel can be simultaneously used by multiple non-conflicting buyers. We also assume that there is a set $\mathbb{N} = \{1, 2, \ldots, n\}$ of buyers. Each buyer $i \in \mathbb{N}$ may request for a single or multiple channel, and has a per-channel valuation v_i. The channel valuation can be the revenue got by the buyer for serving her subscribers. The per-channel valuation v_i is private information to the buyer i herself. It is also known as *type* in the literature. In the auction, the buyers simultaneously submit their sealed bids, denoted by $\mathbf{b} = \{b_1, b_2, \ldots, b_n\}$, which are based on their types. The auction mechanism determines the set of winning buyers, channel allocation to the winners, and the charge of each winner. Denote the charge of a buyer $i \in \mathbb{N}$ by p_i. Then we define the utility u_i of buyer i to be the difference between her valuation v_i on the channel and the charge p_i:

$$u_i = v_i - p_i. \tag{5.1}$$

We assume that the buyers are rational. The objective of each buyer is to maximize her own utility. A buyer has no preference over different outcomes, if the utilities are same to the buyer herself. We also assume that the buyers do not collude with each other.

In contrast to players' individual objective, the overall objective of the auction mechanism is to improve channel utilization. Here, channel utilization is the sum of allocated channels of all the winning buyers.

5.2   Auction Mechanism for Homogenous Channel Allocation

In this section, we present a graph coloring-based auction mechanism for channel allocation. The basic idea of this auction mechanism is to apply existing graph coloring algorithm to a conflict graph, which is created according to the interference among the buyers. Thus, the buyers with the same color are free of interference between each other, and can be allocated on the same channel. The group of buyers with the same color can then be treated as a super buyer. Consequently, the auction can be performed between the seller and the super buyers.

We present a *S*trategy-proof *M*echanism for radio spectrum *ALL*ocation (SMALL). SMALL is a sealed-bid reserve auction mechanism, in which all bidders simultaneously submit sealed bids so that no bidder knows the bid of any of the other participants, and a channel may not be sold if the final bid is not high enough to satisfy the seller.

We consider that the seller has a reserve price for each of the channels, denoted by $s = (s_1, s_2, \ldots, s_m)$. A reserve price can be an operating expense, if the seller put a channel on auction. A channel can be leased to one or a group of non-conflicting buyers if the payment is not lower than the reserve price.

For clarity, we first consider the case that each buyer only requests a single channel, and then extend it to deal with multi-channel bids.

5.2.1   Auction Design with Single-Channel Bids

SMALL is composed of three algorithms: buyer grouping, winner selection, and charge determination. If the seller is a trustworthy authority, we can let the seller serves as auctioneer and perform the computation of the three algorithms. Otherwise, a trusted third party is needed to serve as the auctioneer.

5.2.1.1   Buyer Grouping

Since the channels can be spatially reused, SMALL divides the buyers into multiple non-conflicting groups, each of which can be assigned to a distinguished channel. To prevent the buyers manipulating the auction, the grouping need to be independent of the buyers' bids. Therefore, SMALL first constructs a conflict graph of the buyers. Any pair of buyers, who are in the interference range of each other, have an edge connecting them in the conflict graph. Then buyer groups can be calculated by any existing graph coloring algorithm [64], which is independent of buyers' bids, such that no buyer can be in multiple groups. We note that the buyers cannot determine which group they are in by themselves, when the above grouping strategy is used. We denote the calculated buyer groups by $\mathbb{G} = \{g_1, g_2, \ldots, g_q\}$.

Figure 5.1 shows a toy network with 6 buyers (A–F). There are several grouping results, e.g., $g_1 = \{A, D\}$, $g_2 = \{B, E\}$, and $g_3 = \{C, F\}$.

Fig. 5.1 A toy network with 6 buyers (A–F)

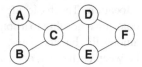

5.2.1.2 Winner Selection

We now determine an integrated group bid for each buyer group. A natural way to calculate the group bid is to simply add all the bids from the group members together. However, this way may allow the buyers to manipulate the group bid by reporting untruthful bids. Thus, the strategy-proofness of the auction can be hurt. Therefore, to guarantee the strategy-proofness, we sacrifice the buyer with the smallest bid in each group, and define an integrated group bid σ_j for each group $g_j \in \mathbb{G}$ as:

$$\sigma_j = (|g_j| - 1) \times \min\{b_k | k \in g_j\}. \tag{5.2}$$

By this way, the group bid is independent of valid members' bids (i.e., the bids except the smallest one) in each group. Such a definition of group bid is reasonable, because the strategy-proofness can be guaranteed by sacrificing the buyer that makes the least contribution in a group. Then, we get a set of group bids $\Sigma = \{\sigma_1, \sigma_2, \ldots, \sigma_q\}$.

Next, SMALL sorts the channels by reserve price in non-decreasing order and buyer groups by group bid in non-increasing order:

$$\mathbb{C}' : c_1', c_2', \ldots, c_m', \quad s.t., \ s_1' \le s_2' \le \ldots \le s_m', \tag{5.3}$$

$$\mathbb{G}' : g_1', g_2', \ldots, g_q', \quad s.t., \ \sigma_1' \ge \sigma_2' \ge \ldots \ge \sigma_q'. \tag{5.4}$$

Here each s_i' (σ_j') corresponds to a unique reserve price in $\mathbf{s}$ (group bid in Σ). In the case of ties, the ordering is random, with each tied channel/group having an equal probability of being ordered prior to the other one.

Next, SMALL finds the maximal number of trades k, s.t.,

$$\sum_{i=1}^{k} s_i' \le \sum_{i=1}^{k} \sigma_i'. \tag{5.5}$$

Finally, the winning groups are the first k buyer groups in $\mathbb{G}'$, and the first k channels in $\mathbb{C}'$ are leased to each of the corresponding winning groups. In each of the winning groups, the buyers, except the one with the smallest bid in that group, are winning buyers. In the case of ties, i.e., more than one buyers report the smallest bid in the group, each tied buyer has an equal probability of being selected as a winning buyer.[1]

Noting that exactly one buyer must be sacrificed for each channel leased, the total number of sacrificed buyers has an upper bound m, which is the number of channels. Since singleton groups cannot compete for channels, as their group bid would be zero, SMALL is more appropriate to be used in a radio spectrum auction with relatively large number of buyers scattered in a large area.

5.2.1.3 Charging

Each winning buyer $i \in g_j$ is charged an even share of her group bid, which is also equivalent to the smallest bid in the group:

$$p_i = \frac{\sigma_j}{|g_j| - 1} = \min\{b_k | k \in g_j\}.$$

In each winning group, we exclude the buyer with the smallest bid, and charge the others with the smallest bid, in order to make the charge be independent of winners' bids.

The seller collects all the payments:

$$P = \sum_{j=1}^{k} \sigma'_j. \tag{5.6}$$

Combining Eqs. (5.5) and (5.6), we get

$$P \geq \sum_{i=1}^{k} s'_i. \tag{5.7}$$

Therefore, the seller's profit is guaranteed. We note that we do not specify the algorithm for dividing the seller's revenue to each channel successfully leased. One of the possible ways is to divide the revenue proportionally to the channels' reserve prices.

Next, we show that SMALL is a strategy-proof mechanism. Before proving the strategy-proofness, we prove the following lemma:

---

[1]We have to note that SMALL is designed to guarantee the truthfulness of channel auction. However, it does not provide any guarantee on the optimality of the channel allocation result. The optimality of the channel allocation result relies on the output of the graph coloring algorithm. We left the problem of selecting the most suitable graph coloring algorithm as an open question.

Lemma 5.1. *If SMALL is used, reporting the true channel valuation as a bid is a dominant strategy for each buyer.*

Proof. We will show that a buyer can not increase her utility by proposing a bid other than her true valuation. That is to say, truthfulness is a dominant strategy.

Consider a buyer i in group g_j with valuation v_i. Let $b_{min} = \min\{b_k | k \in g_j\}$. We distinguish two cases:

1. The buyer i is in a winning group, when bidding true valuation, i.e., $b_i = v_i$. Her utility is

$$u_i = \begin{cases} v_i - b_{min} & \text{if } b_i > b_{min}, \\ 0 & \text{if } b_i = b_{min}. \end{cases}$$

Consider the following two cases:

- $b_i > b_{min}$: Buyer i is a winner. Suppose buyer i reports another bid $b_i' \neq b_i$. If she still wins the channel ($b_i' \geq b_{min}$), then buyer i's utility is not changed, since the smallest bid is still b_{min}. If she losses the channel ($b_i' \leq b_{min}$), then buyer i's utility goes to 0. Therefore, buyer i's new utility $u_i' \leq u_i$.
- $b_i = b_{min}$: Suppose buyer i reports a another bid $b_i' \neq b_i$. If she wins a channel ($b_i' \geq \min\{b_k | k \in g_j \wedge k \neq i\}$) and pay $\min\{b_k | k \in g_j \wedge k \neq i\}$, then her utility becomes

$$u_i' = v_i - \min\{b_k | k \in g_j \wedge k \neq i\}$$
$$\leq v_i - b_{min}$$
$$\leq v_i - b_i$$
$$= 0.$$

If she does not win a channel ($b_i' \leq \min\{b_k | k \in g_j \wedge k \neq i\}$), then her utility is still 0. Therefore, $u_i' \leq u_i$. Recall that the buyer has no preference over different outcomes, if the utilities are same. Therefore, she has no incentives to report a higher bid and win the channel.

2. The buyer i is not in a winning group, when bidding true valuation, i.e., $b_i = v_i$. Her utility is

$$u_i = 0.$$

In this case, the only way to change her utility is to make group g_j becomes a winning group, by reporting a higher bid $b_i' > b_i$, when b_i is the lowest bid in group g_j. Suppose group g_j becomes a winning group when the buyer i reports $b_i' > b_{min} = b_i$. If b_i' is still the smallest bid in group g_j, she still cannot get a channel and her utility is $u_i' = 0$. If $b_i' \geq \min\{b_k | k \in g_j \wedge k \neq i\}$, she may get a channel and pay $\min\{b_k | k \in g_j \wedge k \neq i\}$. Then her utility is

$$u'_i = v_i - \min\{b_k | k \in g_j \wedge k \neq i\}$$
$$\leq v_i - \min\{b_k | k \in g_j\}$$
$$\leq v_i - b_i$$
$$= 0.$$

Therefore, bidding the true valuation is a dominant strategy for each buyer. This completes our proof. □

From the analysis above, we get that SMALL satisfies incentive compatibility. On one hand, we can see that each truthful buyer's utility is always ≥ 0. On the other hand, by not taking part in the auction, a buyer cannot get a channel and her utility remains to be 0. So participating is not worse than staying outside, which satisfies the individual rationality.

Since our mechanism satisfies both incentive compatibility and individual rationality, we have the following theorem:

Theorem 5.1. *SMALL is a strategy-proof mechanism.*

5.2.2 Extension to Multi-Channel Bids

We now consider that the buyers also submit their numbers of channels requested together with the bids to the auctioneer. We denote the profile of channel requests by $\mathbf{r} = (r_1, r_2, \ldots, r_n)$. We can extend SMALL to adapt to multi-channel requests by updating the winner selection algorithm.

5.2.2.1 Updated Winner Selection

SMALL determines the winners and channel allocation iteratively. In each iteration, SMALL sorts the remaining channels by the reserve price in a non-decreasing order, and sorts the remaining buyer groups by the group bids in a non-increasing order:

$$\mathbb{C}^t : c_1^t, c_2^t, \ldots, c_{m^t}^t, \quad s.t., \quad s_1^t \leq s_2^t \leq \ldots \leq s_{m^t}^t, \tag{5.8}$$

$$\mathbb{G}^t : g_1^t, g_2^t, \ldots, g_q^t, \quad s.t., \quad \sigma_1^t \geq \sigma_2^t \geq \ldots \geq \sigma_q^t. \tag{5.9}$$

Here, t indicates that this is in the tth iteration, and m^t denotes the current number of remaining channels. Since the number of groups does not change in different iterations, we can use the same q in all the iterations. If two channels have the same reserve price or two groups have the same group bid, the order between them is determined following the alphabetical order.

Next, SMALL finds the maximal possible number of trades k^t in the tth iteration, s.t.

$$\sum_{i=1}^{k^t} s_i^t \leq \sum_{i=1}^{k^t} \sigma_i^t. \tag{5.10}$$

Finally, SMALL selects the first k^t groups in list $\mathbb{G}^t$ as winning buyer groups, and assigns the first k^t channels in list $\mathbb{C}^t$ to the corresponding winner groups. In each winning group, the buyer(s), except the one who bids the smallest in the group, are winning buyers and can get the channel assigned to the group. We denote the set of winning buyers in the tth iteration by

$$W^t = \bigcup_{j=1}^{k^t} \{i | i \in g_j^t \wedge i \neq \underset{l \in g_j^t}{argmin}(b_l)\}. \tag{5.11}$$

For each winning buyer $i \in W^t$, SMALL decreases its number of requested channels r_i by 1. If $r_i = 0$, then buyer i's demand is fully filled, and SMALL removes buyer i from the buyer group g_j she belongs to and updates group g_j's group bid:

$$g_j = g_j \setminus \{i\},$$

$$\sigma_j = \left(|g_j| - 1\right) \times \min\{b_i | i \in g_j\}.$$

SMALL also deletes the channels, which have already been sold, from the set of channels. SMALL repeats the above procedure until no more winner can be generated (i.e., $k^t = 0$).

The pseudo-code of above winner selection and channel allocation is shown by Algorithm 4. In Algorithm 4, function $GROUPING(\mathbb{N})$ is a graph coloring based grouping algorithm, and returns the buyer grouping result.

Let d denote the largest degree in the conflict graph of the buyers. The computation complexity of the greedy graph coloring based buyer grouping algorithm is $O(n + |E|)$, where E is the set of edges, and the number of groups is at most $(d + 1)$. In each iteration, SMALL takes $\max\{O(m \log m, O(d \log d))\}$ time to sort the bids and reserve price and $O(m)$ time to determine the number of good trades. What's more, SMALL can run at most m iterations. Therefore, the overall computation complexity of SMALL is $O(n + |E| + m \cdot \max\{m \log m, d \log d\})$.

5.2.3   Evaluation Results

We implement SMALL based on a greedy graph coloring algorithm [63], and compare its performance with VERITAS [84]. Buyers are randomly distributed in

---

Algorithm 4 Winner determination and channel allocation algorithm

---

Input: A set of channels $\mathbb{C}$, a profile of reserve prices $\mathbf{s}$, the number of channels m, a set of buyers
   $\mathbb{N}$, a profile of bids $\mathbf{b}$, and a profile of channel requests $\mathbf{r}$.
Output: A set of winning buyers $\mathbb{W}$ and a profile of channel allocation $\mathbf{A}$.
1: $\mathbb{W} \leftarrow \varnothing; \mathbf{A} \leftarrow \varnothing^n; t \leftarrow 1; m^t \leftarrow m$.
2: $(\mathbb{G}, q) \leftarrow GROUPING(\mathbb{N})$.
3: **repeat**
4:     **for all** $g_j \in \mathbb{G}$ **do**
5:         $B_j = (|g_j| - 1) \times min\{b_i | i \in g_j\}$.
6:     **end for**
7:     Sort the channels $\mathbb{C}$ by reserve price $\mathbf{s}$ in non-decreasing order: $c_1^t, c_2^t, \ldots, c_{m^t}^t$, s.t., $s_1^t \leq$
       $s_2^t \leq \ldots \leq s_{m^t}^t$.
8:     Sort buyer groups $\mathbb{G}$ by group bid σ in non-increasing order: $g_1^t, g_2^t, \ldots, g_q^t$, s.t., $\sigma_1^t \geq \sigma_2^t \geq$
       $\ldots \geq \sigma_q^t$.
9:     $k^t \leftarrow \underset{k^t \leq \min\{m^t, q\}}{argmax} \left(\sum_{i=1}^{k^t} s_i^t \leq \sum_{i=1}^{k^t} \sigma_i^t \right)$.
10:     $W^t \leftarrow \bigcup_{j=1}^{k^t} \{i | i \in g_j^t \wedge i \neq \underset{l \in g_j^t}{argmin}(b_l)\}$.
11:     $\mathbb{W} \leftarrow \mathbb{W} \cup W^t$.
12:     **for** $j \leftarrow 1$ to k^t **do**
13:         **for all** $i \in g_j^t \setminus \{\underset{l \in g_j^t}{argmin}(b_l)\}$ **do**
14:             $A_i \leftarrow A_i \cup \{c_j^t\}; r_i \leftarrow r_i - 1$.
15:             **if** $r_i = 0$ **then**
16:                 $g_j^t \leftarrow g_j^t \setminus \{i\}$.
17:             **end if**
18:         **end for**
19:         $\mathbb{C} \leftarrow \mathbb{C} \setminus \{c_j^t\}$.
20:     **end for**
21:     $m^{t+1} \leftarrow m^t - k^t; t \leftarrow t + 1$.
22: **until** $k^t = 0$.
23: **return** $\mathbb{W}$ and $\mathbf{A}$.

---

the terrain area of $2,000 \times 2,000$ m by default. The number of buyers varies from 20
to 400. The radio interference range of each node is set to 425 m. The numbers of
channels for leasing can be 5, 10, or 15. In the single-radio simulation, each buyer
only has a single radio and reports one bid; while in the multi-radio simulation,
each buyer is equipped with 3 radios and can bid for up to 3 channels. We assume
that buyers' channel valuations are randomly distributed over $(0, 1]$, and seller's
reserve prices are randomly distributed over $(0, 2]$.[2] All the results on performance
are averaged over 200 runs.

Figure 5.2 shows channel utilizations of SMALL and VERITAS for auctioning
15 channels. Two sets of results are presented. In one set, each buyer only has
a single radio. In the other set, each buyer is equipped with 3 radios. From the

---

[2]The ranges of buyers' channel valuations and seller's reserve prices can be different from the
ones used here. However, the evaluation results of using different ranges are identical. Therefore,
we only show the results for the above ranges in this paper.

Fig. 5.2 Channel utilizations of SMALL and VERITAS for auctioning 15 channels. (**a**) Single radio (**b**) 3 radios

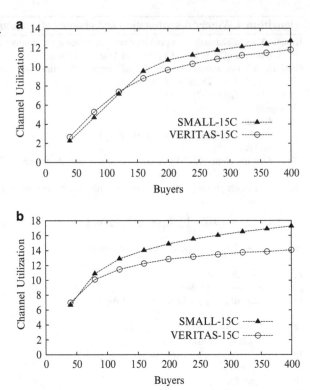

figure, we can see that VERITAS performs better than SMALL when the number of (virtual) buyers is no larger than 120 (In the case of single radio, the threshold is 120 buyers; In the case of 3 radios, the threshold is 40 buyers, which is equal to 120 "virtual" buyers.). This is because VERITAS does not need to sacrifice any bid to guarantee the strategy-proofness. However, when the number of (virtual) buyers is larger than 120, the channel utilization achieved by SMALL becomes higher than that of VERITAS. The reason for this is that VERITAS's greedy channel allocation algorithm is lack of consideration of the whole network. SMALL provides better channel utilization in networks with relatively large number of buyers.

Figure 5.3 shows channel utilizations of SMALL and VERITAS for auctioning 5, 10, and 15 channels, among 200 buyers, when every buyer has a single radio and 3 radios, respectively. The channel utilization of SMALL is higher than that of VERITAS, only except when every buyer has a single radio and there are 5 channels for sale.

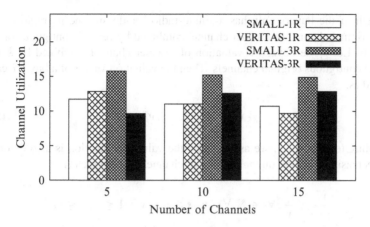

Fig. 5.3 Channel utilizations of SMALL and VERITAS for auctioning 5, 10, and 15 channels among 200 buyers. All the buyers are equipped with 1 or 3 radios

5.3   Auction Mechanism for Adjustable Channel Allocation

In existing papers on channel auction (e.g., [14,65,82,84]), it is commonly assumed that every buyer either bids for only one channel, or bids for multiple channels with the same per-channel price. However, as mentioned in Sect. 4.2, doubling the number of channels, especially contiguous channels, a buyer's valuation does not necessarily double. It has been shown that the saturated throughput is a concave non-decreasing function on channel width [8]. Consequently, according to the saturated throughput on a channel, a buyer's valuation is reasonably expected to be a concave non-decreasing function on the width of the channel she gets. Different buyers may have different valuation functions. Therefore, considering the need for various numbers of channels due to various valuations, it is more reasonable to give the buyers the flexibility to submit various combinatorial bids for contiguous channels.

In this section, we present SPECIAL, which is a *S*trategy-*P*roof and *EffiCI*ent multi-channel *A*uction mechanism for wire*L*ess networks. We model the problem of wireless channel allocation as a *combinatorial channel auction*. Different from existing channel auction mechanisms, our combinatorial channel auction allows buyers to bid for various numbers of contiguous channels. Our model of combinatorial channel auction is a variant of traditional combinatorial auctions, which allow buyers to place bids on any combinations of discrete items.

In our model, the buyers bid for contiguous channels, which can be accessed with a single radio. As is shown in Sect. 4.2, contiguous original channels can be combined to get a wider channel. Such a combined channel normally can provide higher throughput than a single original channel. In our auction, a (combined) channel can be leased to one or a group of non-conflicting buyers. We assume that each of the buyers only has a single radio,[3] and can tune its radio to work on an original channel or a wider channel combined by several contiguous original channels. Let v_i^k be buyer i's valuation of a wider channel combined by k ($1 \leq k \leq m$) contiguous original channels. Then the valuation vector of a buyer i can be denoted as:

$$\mathbf{v}_i = (v_i^1, v_i^2, \ldots, v_i^m). \tag{5.12}$$

According to Sect. 4.2.1, we assume that the valuation function is also a concave non-decreasing function on the number of channels, which means

$$\frac{v_i^x}{x} \geq \frac{v_i^y}{y}, \forall i \in \mathbb{N}, \forall x, y, \ s.t. \ x < y \wedge 1 \leq x, y \leq m, \tag{5.13}$$

In practice, it is more reasonable to give the buyers the flexibility to submit various combinatorial bids for channels. In our combinatorial channel auction, we allow each buyer to submit an independent bid b_i^k for each number k ($1 \leq k \leq m$) of contiguous channels. Similarly, we denote a buyer i's bid vector by:

$$\mathbf{b}_i = (b_i^1, b_i^2, \ldots, b_i^m).$$

According to inequation (5.13), we have

$$\frac{b_i^x}{x} \geq \frac{b_i^y}{y}, \forall i \in \mathbb{N}, \forall x, y, s.t. \ x < y \wedge 1 \leq x, y \leq m, \tag{5.14}$$

when buyers truthfully submit their bids.

5.3.1 Auction Design

The design of SPECIAL is composed of three main components: *buyer grouping and bid integration*, *group-channel allocation*, and *winner selection and charging*.

5.3.1.1 Buyer Grouping and Bid Integration

Considering the spatial reusability of the channels, SPECIAL divides all the buyers into multiple non-conflicting groups. Each group can be assigned with a distinct channel. The assigned channel is either an original channel or a wider channel

---

[3]We note that our channel auction mechanism can be extended to the case of multiple radios by modeling each radio as a virtual buyer [65].

that is composed of several original contiguous channel. To prevent the buyers from manipulating the auction, here we group the buyers using a bid-independent method. As previous section, SPECIAL uses a conflict graph to capture the radio transmission interference among the buyers. Any pair of buyers, who are in the radio transmission interference range of each other, have a line connecting them in the conflict graph. Then the calculation of bid-independent groups can be implemented by a certain existing graph coloring algorithm (e.g., [64]), such that no two buyers have interference between each other in the same group. We note that the buyers have no control on which group they are in, when the above grouping strategy is used.

We denote the set of buyer groups by

$$\mathbb{G} = \{g_1, g_2, \ldots, g_q\}, \tag{5.15}$$

where q is the number of the buyer groups. The buyer groups in $\mathbb{G}$ should satisfy the following two requirements:

$$\bigcup_{1 \le j \le q} g_j = \mathbb{N}, \tag{5.16}$$

meaning that all the buyers are involved, and

$$g_j \cap g_f = \varnothing, \forall g_j, g_f \in \mathbb{G} \wedge j \neq f, \tag{5.17}$$

meaning that no buyer can be in multiple groups.

Fig. 5.4 A toy example with 6 buyers (A–F)

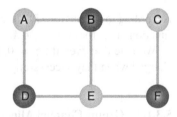

Figure 5.4 shows a toy example with 6 buyers (A–F). There exists several feasible grouping results, e.g., $g_1 = \{A, C, E\}$ and $g_2 = \{B, D, F\}$. This toy example will be further used to illustrate the design rationale of SPECIAL.

From now on, we consider the buyer groups as competitors in the combinatorial channel auction. We now define the integrated group bid for each of the buyer groups. Although a natural way to define the group bid is to simply sum up all the bids for each number of contiguous channels from the group members, this way may allow some of the buyers to manipulate the group bid by reporting untruthful bids [82]. Therefore, to guarantee the strategy-proofness of the auction, we let the group bid be proportional to the smallest bid for each number of contiguous channels in the group, and sacrifice the buyers who may benefit from manipulating the group bid. The sacrificed will not be granted any channel. Two types of buyers have to be sacrificed when computing a group's bid for k contiguous channels:

1. The buyer who submits the smallest bid for k contiguous channels in the group. In the case of ties, i.e., more than one buyer submit the smallest bid in the group, the tied buyer with smallest identification number will be selected as the sacrificed buyer.
2. The buyer who can benefit by manipulating her bid for other numbers of contiguous channels than k in order to make herself win k contiguous channels. In Sect. 5.3.1.3, we will present our scheme to identify such cheating buyers in order to achieve strategy-proofness.

Here, we claim that the number of sacrificed buyers is always no more than two in a buyer group. So we define the integrated group bid (IGB) φ_j^k for each group $g_j \in \mathbb{G}$ on k contiguous channels as

$$\varphi_j^k = \max((|g_j| - 2)\theta_j^k, 0), \tag{5.18}$$

where

$$\theta_j^k = \min_{l \in g_j}(b_l^k). \tag{5.19}$$

We denote the IGB vector of group g_j as

$$\boldsymbol{\varphi}_j = (\varphi_j^1, \varphi_j^2, \ldots, \varphi_j^m). \tag{5.20}$$

According to inequation (5.14), we can get that φ_j^k is also a concave non-decreasing function on k, for every $g_j \in \mathbb{G}$.

We note that even if $\varphi_j^k = 0$, the valid winning buyers in group g_j will still be charged when they successfully win k contiguous channels.

5.3.1.2   Group-Channel Allocation

After forming the buyer groups, we present our algorithm that allocates contiguous channels to the buyer groups based on their IGBs.

For ease of comparison between IGBs, we define per-channel integrated group bid (PIGB) ξ_j^k for each buyer group g_j on k contiguous channels:

$$\xi_j^k = \frac{\varphi_j^k}{k}. \tag{5.21}$$

Algorithm 5 Algorithm for group-channel allocation GCA()

Input: The set of buyer groups $\mathbb{G}$, the number of available channels m, and a set $\xi = \{\xi_j^k | g_j \in \mathbb{G}, 1 \leq k \leq m\}$ of PIGBs.

Output: A vector $\mathbf{r}$ of numbers of channels allocated to every group, and a channel allocation vector $\mathbf{ca}$.

1: $\mathbf{r} \leftarrow 0^q, \mathbf{ca} \leftarrow (0,0)^q, m' \leftarrow m$.
2: **while** $m' > 0$ **do**
3:    $\xi_j^k \leftarrow \max(\xi)$.
4:    $r_j \leftarrow k, m' \leftarrow m' - 1$.
5:    $\xi \leftarrow \xi \setminus \{\xi_j^k\}$.
6: **end while**
7: $m' \leftarrow 1$.
8: **for** $j = 1$ to q **do**
9:    **if** $r_j > 0$ **then**
10:       $ca_j \leftarrow (m', m' + r_j - 1)$.
11:       $m' \leftarrow m' + r_j$.
12:    **end if**
13: **end for**
14: return $(\mathbf{r}, \mathbf{ca})$.

Similarly, we denote the PIGB vector of group g_j as

$$\boldsymbol{\xi}_j = (\xi_j^1, \xi_j^2, \ldots, \xi_j^m). \tag{5.22}$$

Since φ_j^k is a concave non-decreasing function on k, we can get that ξ_j^k is a non-increasing function on k, such that

$$\xi_j^x \geq \xi_j^y, \ \forall x < y \wedge 1 \leq x, y \leq m, \ \forall g_j \in \mathbb{G}. \tag{5.23}$$

For the ease of comparison between PIGBs, we define the preference relations as

$$(a, h) \prec (b, j) \Leftrightarrow a < b \vee (a = b \wedge h < j), \tag{5.24}$$

where a and b are values of PIGBs, and h and j are the identification numbers of buyer groups. In the case of ties in the process of channel allocation, we determine that the group with higher group number has higher priority to be allocated a channel.

Algorithm 5 shows the pseudo-code of group-channel allocation algorithm GCA() used in SPECIAL. The algorithm takes in the set of buyer groups $\mathbb{G}$, the number of available channels m, and a set $\xi = \{\xi_j^k | g_j \in \mathbb{G}, 1 \leq k \leq m\}$ of PIGBs, and then outputs a vector $\mathbf{r}$ of the numbers of original channels allocated to each group, and a vector $\mathbf{wc}$ which determines the channels allocated to every group. Generally, GCA() is a greedy algorithm, and according to inequation (5.23), we can get that, for every group, the number of its allocated channels increases one by one, if any. Therefore, after the execution of the GCA(), there will be no available channel left. Each element r_j in $\mathbf{r}$ means that r_j contiguous channels are allocated to buyer group g_j; each element $wc_j(x, y)$ in **wc** means that the original channels $\{z | x \leq z \leq y\} \subseteq \mathbb{C}$ are allocated to g_j.

We note that although Algorithm 5 can efficiently allocate the channels to the buyer groups according to their PIGBs, it cannot guarantee strategy-proofness. In the next subsection, we will present a method to strengthen Algorithm 5 in order to achieve strategy-proofness.

5.3.1.3   Winner Selection and Charging

In this section, we consider how to determine winners in each winning buyer group who has been assigned channel(s) and their charges for using the assigned channel(s). The design of this part directly determines the auction mechanism's properties. A carefully designed winner selection and charging scheme can guarantee the strategy-proofness of the auction. In this section, we analyze possible cheating actions of the buyers, and then strengthen our winner selection and charging scheme step by step to achieve strategy-proofness.

Our analysis shows that there are three cheating actions, say, *preemptive bidding*, *depreciated bidding*, and *retreat for advancing*, through which a buyer may improve her utility. We provide a method to prevent each of the cheating actions, respectively. In the following, we continue to use the toy example shown in Fig. 5.4 to illustrate the effect of buyers' cheating actions.

5.3.1.4   Preemptive Bidding

The cheating action of preemptive bidding means that a buyer $i \in g_j$ submits a cheating bid vector to make PIGB $\xi_j'^k$ be selected as a winning bid, which would never be selected as a winning group bid if the buyer i bids truthfully. Thus, group g_j wins k channels, and so does i.

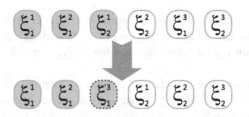

Fig. 5.5 An illustration of preemptive bidding. PIGBs are sorted in non-increasing order. A buyer in group g_1, who can only get 2 channels by bidding truthfully, may get 3 channels by submitting an untruthful bid for 3 channels

Figure 5.5 shows the effect of preemptive bidding. The PIGBs are sorted in non-increasing order. Suppose the sorted PIGBs shown in solid-border round-corner squares represent the case, when buyer $A \in g_1$ bids truthfully. In this case, 2 channels are allocated to g_1 and 1 channel to g_2. Then, we assume that buyer $A \in g_1$ submits an untruthful bid $b'^3_A \neq v^3_A$, such that PIGB ξ^1_1, ξ^2_1, and ξ'^3_1 (indicated by the dashed-border round corner square) are sequentially selected as winning bids. Consequently, group g_1 wins 3 channels, and the buyer A may also get 3 channels.

From this example, we observe that buyer A's truthful bid $b^{\star 3}_A$ must be the minimum bid in $\{b^{\star 3}_A, b^3_C, b^3_E\}$; otherwise, buyer A's cheating bid on 3 channels cannot increase the PIGB ξ^3_1 of group g_1. So $v^3_A = b^{\star 3}_A = \theta^3_1$. If we charge every winner in group g_1 the price θ^q_1, where q channel(s) will be allocated to group g_1, then even if buyer A successfully get 3 channels, her utility will be negative or zero, because θ^3_1 will be at least as large as v^3_A.

Formally, we define the charging scheme as follows. If a group g_j wins k contiguous channels, each potential winning buyer $i \in g_j$ is charged a uniform price, which is equivalent to the smallest bid for k contiguous channel in the group. Here, we define the charge of every buyer for using the allocated channels as

$$p_i = \theta^k_j \eta_i, \tag{5.25}$$

where η_i decides whether buyer i is selected as a winner or not.

Lemma 5.2. *For any winning group bid ξ^k_j, if we charge each winner $i \in g_j$ with θ^k_j, preemptive bidding can be prevented.*

Proof. Suppose $i \in g_j$ submits several cheating bids, among which the smallest is $b'^k_i \neq v^k_i$, such that PIGB ξ^k_j, which is not a winning group bid when i bids truthfully, becomes ξ'^k_j and group g_j is allocated with k channels. Due to Eq. (5.18), (5.19), and (5.21), we can get that $i = \underset{l \in g_j}{argmin}(b^k_l)$, and buyer i submits an untruthful bid $b'^k_i > v^k_i$. Thus, if the buyer i successfully get the channel(s) after cheating, her utility becomes

$$u'_i = v^k_i - \theta' k_j$$
$$\leq v^k_i - b^{\star k}_i$$
$$= 0$$

Therefore, buyer i has no incentive to adopt the cheating action of preemptive bidding. $\square$

5.3.1.5 Depreciated Bidding

The cheating action of depreciated bidding means that a buyer $i \in g_j$ may submit a lower cheating bid b'^k_i than the truthful one $b^{\star k}_i = v^k_i$, with no influence on the channel allocation. Such a cheating action may decrease the charge to the winners in g_j, if ξ'^k_j is a winning PIGB and b'^k_i appears to be the smallest bid for k channels in the group g_j. As a result, if i is selected as an auction winner, her utility can be increased through depreciated bidding.

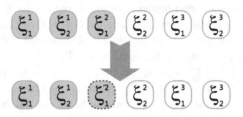

Fig. 5.6 An illustration of depreciated bidding. PIGBs are sorted in non-increasing order. A buyer in group g_1 can increase her utility by submitting an untruthful bid that is lower than any others' bids in the group and her own valuation for 2 channels

Figure 5.6 shows the cheating action of depreciated bidding. Suppose buyer group g_1 wins 2 channels when buyer A bid truthfully (i.e., $b_A^2 = v_A^2$). We assume that buyer A submits a lower bid $b_A'^2 < b_A^{\star 2}$, such that g_1 still wins 2 channels, but $\xi_1'^2 < \xi_1^2$. So $\min(b_A'^2, b_C^2, b_E^2) \leq \min(b_A^{\star 2}, b_C^2, b_E^2)$. Then buyer A's utility becomes

$$u_A' = v_A^2 - \min(b_A'^2, b_C^2, b_E^2)$$

$$\geq v_A^2 - \min(b_A^{\star 2}, b_C^2, b_E^2)$$

$$= u_A.$$

Hence, buyer A may get her utility increased by depreciated bidding.

From this example, we observe that if a buyer $i \in g_j$ can benefit from depreciated bidding, she must appear to be the one who has the smallest bid for k channels when ξ_j^k is a winning PIGB. Therefore, after allocating k channels to buyer group g_j, the buyer, who has the smallest bid for k channels in the group g_j, should be excluded from the set of winners.

Lemma 5.3. *If ξ_j^k is a winning PIGB, we can prevent depreciated bidding by excluding the buyer $i = \arg\min_{i \in g_j}(b_i^k)$ from the winner set. That is, let $\eta_i = 0$.*

Proof. Suppose ξ_j^k is a winning PIGB, and a buyer $i \in g_j$ submits a cheating bid $b_i'^k$, such that $\xi_j'^k$ is still a winning PIGB. So the charge to the winners in g_j is lowered. We can get that $i = \arg\min_{i \in g_j}(b_i^k)$ now. Thus, buyer i is excluded from the winner set, and cannot get any channel. So $u_i = 0$, and buyer i has no incentive to take the cheating action of depreciated bidding. $\square$

5.3.1.6 Retreat for Advancing

The cheating action of retreat for advancing means that if a buyer $i \in g_j$ bids truthfully, PIGB ξ_j^k will be selected as a winning PIGB for group g_j; but if buyer i submits several cheating bids, another PIGB $\xi_j^{k'}$ ($k' < k$) is selected as the final winning PIGB for group g_j. Consequently, buyer i's utility u_i may be increased.

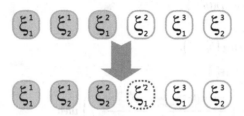

Fig. 5.7 An illustration of retreat for advancing. PIGBs are sorted in non-increasing order. A buyer in group g_1 may get her utility increased, by submitting an untruthful bid for 2 channels to make group g_1 wins 1 channel instead of 2 channels won when bidding truthfully

Figure 5.7 shows the effect of retreat for advancing. If buyer A bids truthfully for 2 channels $b_A^{*2} = v_A^2$, buyer group g_1 will be allocated 2 channels. When buyer A submits a cheating bid $b_A'^2$, only 1 channel is allocated to buyer group g_1. Suppose buyer A is selected as a winner with/without cheating. If $v_A^1 - \min\{b_A^1, b_C^1, b_E^1\} > v_A^2 - \min\{b_A'^2, b_C^2, b_E^2\}$, then the buyer A can get her utility increased by this method of cheating.

From this example, we can observe that if a buyer $i \in g_j$ benefit from winning $k' + 1$ ($\leq k$) channels instead of k through retreat for advancing, then the auction must exhibit the following two properties:

$$i = \underset{l \in g_j}{argmin}(b_l^{k'+1}), \tag{5.26}$$

and

$$\left(\underset{h \neq j}{\min}(\xi_h^{rh}), \underset{h \neq j}{argmin}(\xi_h^{rh}) \right) \prec \left(\frac{\max\left((|g_j| - 2) \cdot \underset{l \neq i \wedge l \in g_j}{\min}(b_l^{k'+1}), 0 \right)}{k' + 1}, j \right) \tag{5.27}$$

To guarantee strategy-proofness, we have to exclude each such buyer who satisfies the above two criteria from the set of winners in each group.

Lemma 5.4. *For every buyer group $g_j \in G$, if there exists buyer $i \in g_j$ satisfying the condition (5.26) and (5.27), we can exclude such buyers from winners to prevent the cheating action of retreating for advancing. That is, let $\eta_i = 0$.*

Algorithm 6 Algorithm for winner selection WIN()

Input: The set of buyer groups $\mathbb{G}$, the number of available channels m, and a set $\xi = \{\xi_j^k | g_j \in \mathbb{G}, 1 \le k \le m\}$.

Output: A set $\mathbb{W}$ of winners in the combinatorial channel auction.

1: $\mathbb{W} \leftarrow \varnothing, pm \leftarrow 0, \xi' \leftarrow \xi$.

2: $(\mathbf{r}, \mathbf{ca}) \leftarrow GCA(\mathbb{G}, m, \xi')$.

3: **for all** $r_j > 0$ **do**

4: $\quad T \leftarrow g_j \setminus \left\{ \underset{i \in g_j}{argmin}(b_i^{r_j}) \right\}$.

5: $\quad$ **if** $r_j < m$ **then**

6: $\quad\quad pm \leftarrow \underset{i \in g_j}{argmin}\left(b_i^{r_j+1}\right)$.

7: $\quad\quad d \leftarrow \underset{h \neq j}{argmin}\left(\xi_h^{r_h}\right)$.

8: $\quad\quad$ **if** $\left(\xi_d^{r_d}, d\right) \prec \left(\dfrac{\max((|g_j|-2) \cdot \underset{l \neq pm \wedge l \in g_j}{\min} (b_l^{r_j+1}), 0)}{r_j+1}, j \right)$ **then**

9: $\quad\quad\quad T \leftarrow T \setminus \{pm\}$.

10: $\quad\quad$ **end if**

11: $\quad$ **end if**

12: $\quad \mathbb{W} \leftarrow \mathbb{W} \cup T$.

13: **end for**

14: **return** $\mathbb{W}$.

Proof. Suppose ξ_j^k is a winning PIGB, and a buyer $i \in g_j$ submits a cheating bid by the method of retreat for advancing, such that $\xi_j^{k'} (k' < k)$ becomes the final winning bid. Thus, we can get $b_i'^{k'+1} = \theta_j'^{k'+1}$. Because if not, $\xi_j'^{k'+1}$ will not be manipulated and buyer group g_j will at least win $k'+1$ channels due to $\xi_j'^{k'+1} \ge \xi_j^k$. Furthermore,

$$\left(\underset{h \neq j}{\min}(\xi_h^{r_h}), \underset{h \neq j}{argmin}(\xi_h^{r_h}) \right) \prec (\xi_j^{k'+1}, l), \tag{5.28}$$

and

$$\xi_j^{k'+1} \le \frac{\max\left((|g_j| - 2) \cdot \underset{l \neq i \wedge l \in g_j}{\min} (b_l^{k'+1}), 0 \right)}{k'+1}. \tag{5.29}$$

So we can get that condition (5.27) holds. Therefore, buyer i is excluded from the winner set, and cannot get any channel. So $u_i = 0$, and thus buyer i has no incentive to take the cheating action of retreat for advancing. $\square$

Finally, we use Algorithm 6 to summarize our method of determining winners. Next, we show the strategy-proofness of SPECIAL.

Theorem 5.2. *SPECIAL is a strategy-proof combinatorial channel auction mechanism.*

Proof. We prove that SPECIAL satisfies both individual rationality and incentive compatibility as follows.

Individual Rationality

We can see that each truthful buyer's utility is always no less than 0. By not taking part in the auction, a buyer cannot get a channel, and her utility is 0. So participating is not worse than staying outside the auction. Therefore, SPECIAL satisfies the individual rationality.

Incentive Compatibility

We will prove that no buyer can increase her utility by submitting a cheating bid, which is not equal to her valuation, no matter what the other buyers do. That is to say, truthful bidding is every buyer's dominant strategy.

Suppose a buyer i belongs to a buyer group g_j that wins r_j and r'_j channels, when buyer i bids truthfully and not, respectively. Here, we note that group g_j wins r_j or r'_j channels does not guarantee that the buyer i also get r_j or r'_j channels, because buyer i can be out of the set of winners. We consider the possible change of buyer i's utility in three cases:

Case 1: $r'_j > r_j$

Group g_j gets more channels when buyer i bids untruthfully. In this case, we can get that

$$b_i'^k > b_i^{*k} = \theta_j^k, \forall k \in \{x | r_j + 1 \le x \le r'_j\}. \tag{5.30}$$

So we can get that

$$b_i'^{r'_j} \ge \theta_j'^{r'_j} > b_i^{*r'_j} = v_i^{r'_j}. \tag{5.31}$$

Thus, if buyer i wins a certain number of channel(s) finally (i.e., $\eta_i = 1$), we can get that

$$u'_i = v_i^{r'_j} - \theta_j'^{r'_j}$$
$$< v_i^{r'_j} - b_i^{r'_j}$$
$$= 0$$

Otherwise, $u'_i = 0$. Therefore, it shows that if buyer i manipulates her bid to win $r'_j > r_j$ channel(s), she will have $u'_i \le u_i$.

Case 2: $r_j' = r_j$

Group g_j still gets the same number of channels when the buyer i bids untruthfully. If $r_j' = r_j = 0$, then buyer i's utility is still 0. So we focus on the cases, in which $r_j' = r_j > 0$. We now distinguish two cases as follows:

- Buyer i wins r_j channels when she submits a truthful bid. In this case, with the fact that $i \neq \underset{l \in g_j}{argmin}(b_l^{r_j})$, her utility is

$$u_i = v_i^{r_j} - \min_{l \in g_j \wedge l \neq i} (b_l^{r_j}). \qquad (5.32)$$

To improve her utility, she has to decrease the charging price for herself. However, she will not manage to reach it unless she decreases her bid $b_i^{r_j}$ to $b_i''^{r_j}$ such that

$$b_i''^{r_j} < \min_{l \in g_j \wedge l \neq i} (b_l^{r_j}). \qquad (5.33)$$

However, if she does so, she will win no channel because of Lemma 5.3, leading to $u_i' \leq u_i$.

- The buyer i wins no channel when she submits a truthful bid, meaning $u_i = 0$ (i.e., $\eta_i = 0$). Due to Lemmas 5.3 and 5.4, which can lead to $\eta_i = 0$, we further distinguish two cases:

  - Using depreciated biding, i.e., $i = \underset{l \in g_j}{argmin}(b_l^{r_j})$. For this case, if she wants to improve her utility, the only possible method is to submit an untruthful bid $b_i''^{r_j} \geq \min_{l \in g_j \wedge l \neq i} (b_l^{r_j})$. But if she does, we will get that

    $$u_i' = v_i^{r_j'} - \min_{l \in g_j \wedge l \neq i} (b_l^{r_j})$$
    $$\leq v_i^{r_j'} - b_i''^{r_j}$$
    $$= 0$$

  - Using retreat for advancing, i.e.,

$$\begin{cases} \left(\underset{h \neq j}{\min(\xi_h^{r_h})}, \underset{h \neq j}{argmin(\xi_h^{r_h})} \right) \prec \left(\dfrac{\max\left((|g_j|-2) \cdot \underset{l \neq i \wedge l \in g_j}{\min} (b_l^{r_j+1}), 0 \right)}{r_j+1}, j \right) \\[4mm] i = \underset{l \in g_j}{argmin} \left(b_l^{r_j+1} \right) \end{cases} \qquad (5.34)$$

For this case, if she wants to improve her utility, the only possible method is to submit an untruthful bid $b_i''^{r_j+1} > b_i^{*r_j+1}$, such that $i \neq \underset{l \in g_j}{argmin} \left(b_l''^{r_j+1} \right)$.

But if she does so, we will get that group j will win $r_j + 1$ channels, because

$$\xi_j''^{r_j+1} = \frac{\max \left((|g_j| - 2) \underset{l \neq i \wedge l \in g_j}{\min} \left(b_l''^{r_j+1} \right), 0 \right)}{r_j + 1} \qquad (5.35)$$

$$\Rightarrow \left(\underset{h \neq j}{\min}(\xi_h^{r_h}), \underset{h \neq j}{argmin}(\xi_h^{r_h}) \right) \prec \left(\xi_j''^{r_j+1}, j \right) \qquad (5.36)$$

However, this contradicts with the condition $r_j' = r_j$. So this cheating method cannot happen in this case.

Therefore, it appears that if buyer i manipulates the auction, achieving that $r_j' = r_j$, we can always have $u_i' \leq u_i$.

Case 3: $r_j' < r_j$

Group g_j gets less channels when buyer i bids untruthfully. If $r_j' = 0$, then buyer i's utility becomes 0. Consequently, we focus on the case of $r_j' > 0$. In this case, we can get that

$$i = \underset{l \in g_j}{argmin} \left(b_l''^{r_j'+1} \right), \qquad (5.37)$$

because otherwise, $\xi_j''^{r_j'+1} = \xi_j'^{r_j'+1} \geq \xi_j'^{r_j}$, leading to the result that group j wins at least $r_j' + 1$ channel(s). We can also get that

$$\frac{\max \left((|g_j| - 2) \underset{l \in g_j \wedge l \neq i}{\min} \left(b_l'^{r_j'+1} \right), 0 \right)}{r_j' + 1} \geq \xi_j'^{r_j'+1} \geq \xi_j'^{r_j} \qquad (5.38)$$

and

$$\left(\underset{h \neq j}{\min}(\xi_h^{r_h}), \underset{h \neq j}{argmin} \left(\xi_h^{r_h} \right) \right) \prec \left(\xi_j'^{r_j}, j \right). \qquad (5.39)$$

Thus, we can conclude that

$$
\left(\min_{h \neq j} \left(\xi_h^{r'_h} \right), \operatorname*{argmin}_{h \neq j} \left(\xi_h^{r'_h} \right) \right) \prec \left(\frac{\max \left((|g_j| - 2) \min_{l \in g_j \wedge l \neq i} \left(b_l^{r'_j + 1} \right), 0 \right)}{r'_j + 1}, j \right)
$$
(5.40)

Hence, according to Lemma 5.4, buyer i will be excluded from the winner set, which results in that $u'_i = 0$.

Therefore, if buyer i manipulates her bid to win $r'_j < r_j$ channel(s), her utility $u'_i \leq u_i$.

All in all, we have proved that truthful bidding is every buyer's dominant strategy. Therefore, SPECIAL satisfies incentive compatibility. Since SPECIAL satisfies both incentive compatibility and individual rationality, we conclude that SPECIAL is a strategy-proof combinatorial channel auction mechanism.   □

5.3.2   Evaluation Results

We implement SPECIAL based on a greedy graph coloring algorithm [63] and present its performance. In the simulation, buyers are randomly distributed in various terrain areas (including $1,000 \times 1,000$, $1,500 \times 1,500$, $2,000 \times 2,000$, and $2,500 \times 2,500$ m). The number of buyers varies from 20 to 600. The radio interference range of each node is set to 425 m. The numbers of channels for leasing can be 6, 12, or 24.

We assume that the vector of any buyer's channel valuations is concave non-decreasing and randomly distributed over $[0, m]$, where m denotes the number of channels.[4] In our simulations, we define the function of the valuation of each buyer as

$$
\begin{cases}
y_1 = \Delta_1 = rand\,[0, 1]; \\
\Delta_t = \Delta_{t-1} * rand\,[0, 1], \ \forall t, \ 2 \leq t \leq m; \\
y_t = y_{t-1} + \Delta_t, \ \forall t, \ 2 \leq t \leq m
\end{cases}
$$
(5.41)

All the results on performance are averaged over 4,000 runs.

We compare the performance of SPECIAL, in terms of channel utilization, with three existing strategy-proof channel auction mechanisms: TRUST [82], VERITAS [84], and SMALL [65].

---

[4]The ranges of buyers' channel valuations can be different from the ones used here. However,the evaluation results of using different ranges are identical. Therefore, we only show the results for the above ranges in this paper.

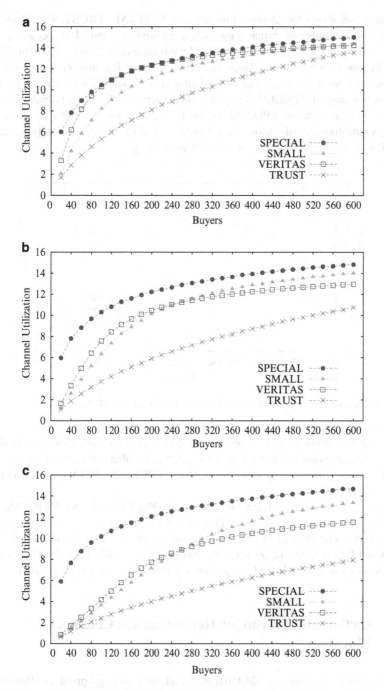

Fig. 5.8 Channel Utilization of SPECIAL, TRUST, VERITAS, and SMALL for auctioning 6, 12, and 24 channels. (**a**) 6 channels (**b**) 12 channels (**c**) 24 channels

Figure 5.8 shows the channel utilization of SPECIAL, TRUST, VERITAS and SMALL. The number of channels for auctioning are 6, 12, and 24, respectively. The number of buyers varies from 20 to 600. The terrain area is 2,000 × 2,000 m. These three figures show that SPECIAL performs at least as well as any other one with any number of buyers involved. The only chance that VERITAS's channel utilization approaches that of SPECIAL happens when there are 6 channels and 120–240 buyers. This is because SPECIAL needs to sacrifice some buyers to guarantee strategy-proofness. Furthermore, the advantage of SPECIAL dramatically increases with the increasing number of channels for auctioning.

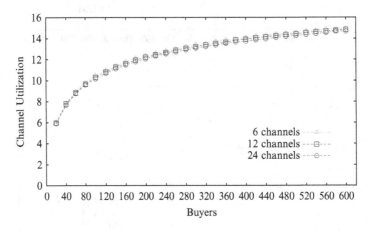

Fig. 5.9 Channel Utilization of SPECIAL for auctioning 6, 12, and 24 channels

Figure 5.9 shows an unique property of SPECIAL, i.e., SPECIAL can achieve almost the same channel utilization with different numbers of channel for auctioning. However, this property does not hold for TRUST, VERITAS, or SMALL as shown in Fig. 5.8.

Figure 5.10 shows the channel utilization of SPECIAL, TRUST, VERITAS, and SMALL with the constant density of buyers over terrain area. We fix the density on $1/20,000\,\mathrm{m}^2$. The number of channels for auctioning are 12. while the terrain area varies as 1,000 × 1,000, 1,500 × 1,500, 2,000 × 2,000, and 2,500 × 2,500 m. Again, the results show that SPECIAL outperforms all the other auction mechanisms.

5.4   Auction Mechanism for Heterogeneous Channel Allocation

In this section, we present SMASHER, which is a *S*trategy-proof co*M*binatorial *A*uction mechani*S*ms for *HE*terogeneous channel *R*edistribution. SMASHER achieves both strategy-proofness and approximately efficient social welfare. Here, social welfare is defined as follows.

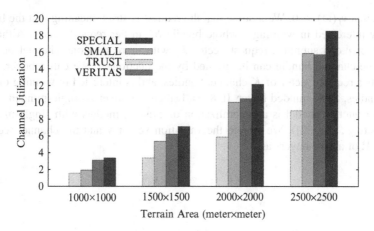

Fig. 5.10 Channel utilization of SPECIAL, TRUST, VERITAS, and SMALL with the constant density of buyers over terrain area. The density is $1/20,000\,\text{m}^2$, and the number of channels for auctioning is 12

Definition 5.1 (Social Welfare). The social welfare in a channel auction is the sum of winning buyers' valuations on their allocated bundles of channels.

$$SW \triangleq \sum_{i \in \mathbb{W}} v_i, \qquad (5.42)$$

where $\mathbb{W}$ is the set of winners.

In contrast to the auction model in Sect. 5.1, we consider that the channels for leasing are *heterogeneous*, and thus the buyers have their own preference over the channels due to spatial variance (e.g., background noise, temperature, and landform). Since wireless devices can be equipped with multiple radios, the buyers may request more than one channel according to their requirements of QoS. Considering the diversity of QoS demand and heterogeneity of the channels, we allow the buyers to submit multiple channel requests, among which one of the requests can be granted. We assume that the buyers have uniform valuation over any of their channel requests, because the buyers' requirement of QoS can be satisfied if one of their requested bundles is allocated.

Each buyer $i \in \mathbb{N}$ submits a vector of requested channel bundles

$$R_i \triangleq \left(S_i^1, S_i^2, \dots, S_i^K \right) \qquad (5.43)$$

to the auctioneer. Any channel bundle $S_i^j \subseteq \mathbb{C}, 1 \le j \le K$ can satisfy her QoS. Each buyer $i \in \mathbb{N}$ has a uniform valuation v_i over any requested channel bundles in R_i. The buyer valuation has two properties: *Free Disposal* and *Normalization*. Free disposal means that for any two subsets of channels S and T, if $S \subseteq T$, then $v_i(S) \le v_i(T)$ (actually, $v_i(S) = v_i(T)$ in our model); while normalization means that $v_i(\varnothing) = 0$. We assume that the request is strict, meaning that the buyer is only interested in winning a whole bundle S_i^j in her request vector. Although the buyer i can submit a request vector R_i with more than one channel bundle, only one channel bundle can be granted by the auctioneer. We call a buyer, who submits a request vector of K channel bundles, and is interested in winning one of the bundles, as K-minded buyer. If $K = 1$, then the buyer is single-minded. Note that our auction model is a generalization of existing models with single-minded buyers (e.g., [17, 22]). We denote the valuation vector $\mathbf{v}$ and the channel request vector $\mathbf{R}$ of all the buyers as

$$\mathbf{v} \triangleq (v_1, v_2, \ldots, v_n), \tag{5.44}$$

and

$$\mathbf{R} \triangleq (R_1, R_2, \ldots, R_n), \tag{5.45}$$

respectively.

Each buyer $i \in \mathbb{N}$ submits a bid b_i to the auctioneer, meaning that if she wins any channel bundle $S_i^j \in R_i$, she would like to pay no more than b_i for it. Here, the bid b_i may not necessarily be equal to her valuation v_i. Let vector $\mathbf{b}$ represent the bids of all the buyers $\mathbf{b} \triangleq (b_1, b_2, \ldots, b_n)$.

5.4.1 Model of Multi-Unit Combinatorial Channel Auction

Different from previous sections, we introduce the concept of *virtual channel* to represent the confliction of channel usage among the buyers. By introducing virtual channels, we transform the problem of heterogeneous channel allocation to a classic multi-unit combinatorial auction, which is computationally intractable. Therefore, we present a strategy-proof and approximately efficient combinatorial auction mechanism for heterogeneous channel redistribution.

5.4.1.1 Virtual Channel

We introduce *virtual channel* to capture the interference among the buyers on different channels. Specifically, a virtual channel $vc_{i,j}^k$ denotes that the buyer i and the buyer j may cause interference between each other on channel c_k, and thus they cannot work on channel c_k simultaneously. Since virtual channel $vc_{i,j}^k$ represents the exclusive usage of channel c_k between the buyer i and j, its quantity is set to 1. When virtual channel $vc_{i,j}^k$ is added to the requested bundle(s) that contains channel c_k from the buyer i and j, at most one of the requests containing channel c_k from the two buyers can be granted. Consequently, the exclusive usage of channel c_k

---

Algorithm 7 Virtual channel generation

Input: A set of conflict graph $\mathbb{G}$, a vector of channel requests $\mathbf{R}$.
Output: A set of virtual channels $\mathbb{VC}$, a vector of updated requests $\mathbf{R}'$.
1: $\mathbb{VC} \leftarrow \varnothing$; $\mathbf{R}' \leftarrow \mathbf{R}$.
2: **for all** $\mathbb{G}_k = (\mathbb{V}_k, \mathbb{E}_k) \in \mathbb{G}$ **do**
3:     **for all** $(i, j) \in \mathbb{E}_k$ **do**
4:         Create virtual channel $vc_{i,j}^k$.
5:         $\mathbb{VC} \leftarrow \mathbb{VC} \cup \left\{ vc_{i,j}^k \right\}$.
6:         **for all** $S_i''^l \in R_i'$, s.t., $c_k \in S_i''^l \wedge \left(\exists S_j''^l \in R_j', c_k \in S_j''^l \right)$ **do**
7:             $S_i''^l \leftarrow S_i''^l \cup \left\{ vc_{i,j}^k \right\}$.
8:         **end for**
9:         **for all** $S_j''^l \in R_j'$, s.t., $c_k \in S_j''^l \wedge \left(\exists S_i''^l \in R_i', c_k \in S_i''^l \right)$ **do**
10:           $S_j''^l \leftarrow S_j''^l \cup \left\{ vc_{i,j}^k \right\}$.
11:         **end for**
12:     **end for**
13: **end for**
14: **return** $\mathbb{VC}$ and $\mathbf{R}'$.

---

between the buyer i and j is guaranteed. The channel redistribution problem can be converted to the allocation of exclusive virtual channels. We present the definition of virtual channel as follows.

Definition 5.2 (Virtual Channel). There is a virtual channel $vc_{i,j}^k$, if the buyer i and buyer j are within the interference range of each other on channel c_k.

In most of existing works on channel auction, a single conflict graph is used to represent the interference among buyers. However, in case of heterogeneous channels, each channel may have a distinctive conflict graph. Let $\mathbb{G}_k \triangleq (\mathbb{V}_k, \mathbb{E}_k)$ denote the conflict graph on channel c_k, where $\mathbb{V}_k \subseteq \mathbb{N}$ is the set of buyers who can access channel c_k, and each edge $(i, j) \in \mathbb{E}_k$ represents the interference between the buyer i and j on channel c_k. Let $\mathbb{G} \triangleq \{\mathbb{G}_k | c_k \in \mathbb{C}\}$ denote the set of conflict graphs.

Since the conflict graph is commonly assumed to be available in wireless networks, we construct the virtual channel from the conflict graph. The process of converting the edges in the conflict graphs to virtual channels with unit quantity is shown by Algorithm 7. Let $\mathbb{VC}$ be the set of virtual channels. We create a virtual channel $vc_{i,j}^k$ (line 4), if there is an edge between the buyer i and j in conflict graph $\mathbb{G}_k$, and append $vc_{i,j}^k$ to the requested bundle(s) containing channel c_k from the buyer i and j, while remaining the corresponding bid(s) unchanged (lines 6–11).

Theorem 5.3. *The number of virtual channels generated has an upper bound* $mn(n-1)/2$.

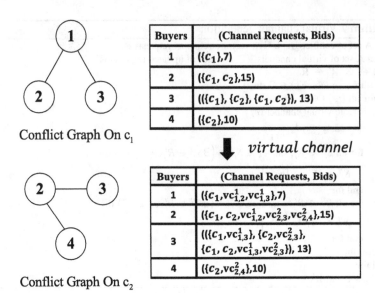

Buyers	(Channel Requests, Bids)
1	$(\{c_1\},7)$
2	$(\{c_1, c_2\},15)$
3	$(\{\{c_1\}, \{c_2\}, \{c_1, c_2\}\}, 13)$
4	$(\{c_2\},10)$

Conflict Graph On c_1

⬇ *virtual channel*

Buyers	(Channel Requests, Bids)
1	$(\{c_1, vc_{1,2}^1, vc_{1,3}^1\},7)$
2	$(\{c_1, c_2, vc_{1,2}^1, vc_{2,3}^2, vc_{2,4}^2\},15)$
3	$(\{\{c_1, vc_{1,3}^1\}, \{c_2, vc_{2,3}^2\}, \{c_1, c_2, vc_{1,3}^1, vc_{2,3}^2\}\}, 13)$
4	$(\{c_2, vc_{2,4}^2\},10)$

Conflict Graph On c_2

Fig. 5.11 An example showing the generation of virtual channels

Proof. In the worst case, every pair of buyers are in the interference range of each other on every channel. Then the maximum number of virtual channels generated from one conflict graph is $n(n-1)/2$. Since there are m conflict graphs, the maximum number of virtual channels is $mn(n-1)/2$.   □

We use a simple example in Fig. 5.11 to explain the concept of virtual channel. In Fig. 5.11, there are 2 channels and 4 buyers. The two conflict graphs show the interference among buyers on two heterogeneous channels c_1 and c_2. The upper right table shows the buyers' channel demands. Both single-minded and multi-minded buyers exist in this example. Here, the buyer 2 is a single-minded buyer, and only bids a bundle of channels $(\{c_1, c_2\})$ for 15; the buyer 3 is a multi-minded buyer, and submits three requests, including $(\{c_1\}, \{c_2\}, \{c_1, c_2\})$, and an uniform valuation 13. After executing Algorithm 7, the updated request vectors with virtual channels are shown in the lower right table. Let's see buyer 2's updated request as an example. Since both buyer 1 and buyer 2 bid for channel c_1 and they interfere with each other on this channel, we add a virtual channel $vc_{1,2}^1$ with unit quantity to the buyer 2's single requested bundle.

5.4.1.2 Multi-Unit Combinatorial Channel Auction

Given the virtual channel introduced in last section, we are ready to transform the problem of heterogenous channel allocation to a classic multi-unit combinatorial auction. The outcome of the auction is the set of winning buyers and their assigned channel set.

The goods in the multi-unit combinatorial auction are the channels and virtual channels. The quantities of each channel $c_k \in \mathbb{C}$ and virtual channel $vc_{i,j}^k \in \mathbb{VC}$ are n and 1, respectively. Given the vector of requests with virtual channels $\mathbf{R}'$ and the bid vector $\mathbf{b}$, the auctioneer determines the winners and which channel bundles to grant. Let $x\left(i, S_i^{\prime j}\right) = 1$ denote that the channel set $S_i^{\prime j}$ is granted to the buyer i; otherwise, $x\left(i, S_i^{\prime j}\right) = 0$. The process of winner determination can be modeled as a binary program. The objective is to maximize the social welfare. We use b_i, instead of v_i, because the strategy-proof mechanisms shown in later sections will guarantee that bidding truthfully is the dominate strategy of each buyer $i \in \mathbb{N}$.

Objective:

$$Maximize \quad \sum_{i \in \mathbb{N}} \sum_{j=1}^{K} x(i, S_i^{\prime j}) \times b_i$$

Subject to:

$$\sum_{i \in \mathbb{N}} \sum_{S_i^{\prime j} \in \mathbf{R}'_i, S_i^{\prime j} \ni c_k} x\left(i, S_i^{\prime j}\right) \leq n \qquad \forall c_k \in \mathbb{C} \qquad (5.46)$$

$$\sum_{i \in \mathbb{N}} \sum_{S_i^{\prime j} \in \mathbf{R}'_i, S_i^{\prime j} \ni vc_k} x\left(i, S_i^{\prime j}\right) \leq 1 \qquad \forall vc_k \in \mathbb{VC} \qquad (5.47)$$

$$\sum_{j=1}^{K} x\left(i, S_i^{\prime j}\right) \leq 1 \qquad \forall i \in \mathbb{N} \qquad (5.48)$$

$$x\left(i, S_i^{\prime j}\right) \in \{0, 1\} \qquad \forall i \in \mathbb{N}, 1 \leq j \leq K \qquad (5.49)$$

Here constraints (5.46) and (5.47) indicate the quantity limitation of the channel and the virtual channel, respectively. Constraint (5.48) indicates that each buyer can win at most one bundle of channels out of her submitted requests. Constraint (5.49) indicates the binary value of the auctioneer's decision of allocation.

If the optimal social welfare can be achieved by solving the above binary program, then the celebrated VCG mechanism (named after Vickrey [58], Clark [12], and Groves [28]) can be applied to calculate the clearing price that can ensure the strategy-proofness of the auction mechanism. Unfortunately, the above winner determination problem can be proven to be NP-hard by reducing to the *exact cover* problem [24]. Considering the computational intractability of the winner determination problem, we present an alternative solution with greedy channel allocation to achieve approximately efficient social welfare in the next section. Furthermore, we integrate the greedy allocation algorithm with a novel pricing mechanism to provide a strategy-proof and approximately efficient combinatorial auction mechanism for heterogeneous channel redistribution.

5.4.2 Auction Design

We consider the case of indivisible channels, which can only be allocated exclusively to non-interfering buyers, in this section. As shown in Sect. 5.4.1.2, finding the optimal auction decision is computationally intractable. Existing works [16, 39] show that it is impossible to design a strategy-proof approximation combinatorial auction mechanism in the general case, even if the goods are not spatially reusable.

In this section, we present the design details of SMASHER. SMASHER consists of the following three major components: virtual channel generation, winner determination, and clearing price calculation.

5.4.2.1 Virtual Channel Generation

The process of virtual channel generation is the same as that of Algorithm 7 shown in Sect. 5.4.1.1, except that we add one more virtual channel vc_i with unit quantity to each requested bundle of buyer $i \in \mathbb{N}$. Virtual channel vc_i is used to ensure that at most one of the requested bundles from the buyer i can be granted.

$$S_i'^j = S_i'^j \cup \{vc_i\}, i \in \mathbb{N}, 1 \le j \le K, \tag{5.50}$$

where $S_i'^j$ is the updated bundle with virtual channels. The set of virtual channels is also updated

$$\mathbb{VC} = \mathbb{VC} \cup \{vc_i | i \in \mathbb{N}\}. \tag{5.51}$$

5.4.2.2 Winner Determination

Before presenting the approximation algorithm for winner determination, we introduce *virtual bid*. The uniform virtual bid $\tilde{b}_i$ over any of requested bundles from the buyer i is defined as

$$\tilde{b}_i \triangleq \frac{b_i}{\max_{1 \le l \le K} \left(\sqrt{|S_i'^l|} \right)}. \tag{5.52}$$

SMASHER sorts all the buyers according to their virtual bids in non-increasing order:

$$\mathbb{L}_1 : \tilde{b}_1 \ge \tilde{b}_2 \ge \ldots \ge \tilde{b}_n. \tag{5.53}$$

In case of a tie, SMASHER breaks the tie following a bid-independent rule, such as lexicographic order of buyers' ID and channel number.

Algorithm 8 Approximation algorithm for winner determination

Input: Vector of updated channel requests $\mathbf{R}'$, vector of bids $\mathbf{b}$.
Output: A pair of sets of winning buyers and allocated bundles of channels $(\mathbb{W}, \mathbb{S})$.
1: $(\mathbb{W}, \mathbb{S}) \leftarrow (\emptyset, \emptyset)$.
2: $\mathbb{V} \leftarrow \emptyset$.
3: **for all** $i \in \mathbb{N}$ **do**
4: $\quad \tilde{b}_i \leftarrow b_i / \max_{1 \leq l \leq K} \left(\sqrt{|S_i'^l|} \right)$.
5: **end for**
6: Sort $\tilde{b}_i$ in non-increasing order $\mathbb{L}_1 : \tilde{b}_1 \geq \tilde{b}_2 \geq \ldots \geq \tilde{b}_n$.
7: **for all** $i = 1, \ldots, n$ **do**
8: $\quad$ Sort $S_i'^j$ in non-decreasing order of bundle size $\mathbb{L}_2 : |S_i'^1| \leq |S_i'^2| \leq \ldots \leq |S_i'^K|$.
9: $\quad$ **for all** $j = 1, \ldots, K$ **do**
10: $\quad\quad$ **if** $S_i'^j \cap \mathbb{V} = \emptyset$ **then**
11: $\quad\quad\quad$ Add virtual channels in $S_i'^j$ to $\mathbb{V}$.
12: $\quad\quad\quad$ $(\mathbb{W}, \mathbb{S}) \leftarrow \left(\mathbb{W} \cup \{i\}, \mathbb{S} \cup \{S_i'^j\} \right)$.
13: $\quad\quad\quad$ **break.**
14: $\quad\quad$ **end if**
15: $\quad$ **end for**
16: **end for**
17: **return** $(\mathbb{W}, \mathbb{S})$.

Following the order in $\mathbb{L}$, SMASHER greedily grants the smallest channel bundle, in which no virtual channel has already been allocated, to each buyer.

Algorithm 8 shows the pseudo-code of above winner determination process. In practice, the number of buyers n is much larger than K, thus the complexity of Algorithm 8 is $O(n \log n)$.

5.4.2.3 Clearing Price Calculation

The clearing price is calculated based on *critical virtual bid*.

Definition 5.3 (Critical Virtual Bid). The critical virtual bid $cr(i) \in \mathbb{L}_1$ of buyer $i \in \mathbb{N}$ is the largest virtual bid, after which has been selected as winning virtual bid by Algorithm 8 given the other buyers' requests and bids $(\mathbf{R}'_{-i}, \mathbf{b}_{-i})$, such that none of buyer i's requests can be satisfied.

We note that according to the definition of critical virtual bid, no matter which request of the buyer i is granted in the auction, the critical virtual bid $cr(i)$ is always the same.

We now show the method of calculating the clearing price of the buyer i by distinguishing two cases:

1. If the buyer i loses in the auction or $cr(i)$ does not exist (denoted by $cr(i) = 0$), then her clearing price is 0.

2. If the buyer i is granted channel bundle $\hat{S}_i'^j$ and there exists a critical virtual bid $cr(i)$, the clearing price p_i of buyer i is set to

$$p_i \triangleq cr(i) \times \max_{1 \le l \le K} \left(\sqrt{|S_i'^l|} \right). \tag{5.54}$$

We next prove the strategy-proofness and analyze the approximation ratio of SMASHER.

5.4.2.4   Strategy-Proofness

Theorem 5.4. *SMASHER-AP is a strategy-proof combinatorial auction mechanism for heterogeneous indivisible channel redistribution.*

Proof. We first show that buyer $i \in \mathbb{N}$ cannot obtain higher utility by bidding untruthfully.

We distinguish two cases:

- Buyer i wins bundle $\hat{S}_i'^j$ and gets utility $u_i \ge 0$ when bidding truthfully, i.e., $b_i = v_i$. Let $\hat{S}_i'^j \ne \hat{S}_i'^j$ be the bundle won by the buyer i, when she cheats the bid, i.e., $b'_i \ne v_i$. The utility of buyer i remains the same:

$$u'_i = v_i - p'_i$$

$$= v_i - cr(i) \times \max_{1 \le l \le K} \left(\sqrt{|S_i'^l|} \right)$$

$$= u_i.$$

  If buyer i loses the auction when she cheats the bid, her utility is 0, which is not better than that gained when bidding truthfully.
- The buyer i loses in the auction when bidding truthfully. Then, her utility $u_i = 0$. If she still loses when bidding untruthfully, her utility cannot be changed. We consider the case, in which she cheats the bid $b'_i \ne v_i$ and wins a bundle $\tilde{S}_i'^j \ne \varnothing$. We denote virtual bid $\tilde{b}'_i$ and $\tilde{b}_i$ for channel bundle $\hat{S}_i'^j$ when the buyer i bids truthfully and untruthfully, respectively. Then, we have $\tilde{b}'_i \ge cr(i) \ge \tilde{b}_i$, because otherwise, she still cannot win any bundle. Her utility now becomes non-positive:

$$u'_i = v_i - p'_i$$

$$= v_i - cr(i) \times \max_{1 \le l \le K} \left(\sqrt{|S_i'^l|} \right)$$

$$\le v_i - \tilde{b}_i \times \max_{1 \le l \le K} \left(\sqrt{|S_i'^l|} \right)$$

$$= v_i - \frac{v_i}{\max_{1 \le l \le K} \left(\sqrt{|S_i''|} \right)} \times \max_{1 \le l \le K} \left(\sqrt{|S_i''|} \right)$$

$$= v_i - v_i$$

$$= 0.$$

The buyer i cannot increase her utility by bidding any other value than v_i. Bidding truthfully is a dominant strategy for each buyer. Therefore, SMASHER satisfies incentive compatibility.

We now prove that SMASHER also satisfies individual rationality. On one hand, buyer i's utility is zero if she loses in the auction. On the other hand, the winning buyer i gets utility:

$$u_i = v_i - p_i$$

$$= v_i - cr(i) \times \max_{1 \le l \le K} \left(\sqrt{|S_i^l|} \right)$$

$$= \left(\frac{v_i}{\max_{1 \le l \le K} \left(\sqrt{|S_i^l|} \right)} - cr(i) \right) \times \max_{1 \le l \le K} \left(\sqrt{|S_i^l|} \right)$$

$$= \left(\tilde{b}_i - cr(i) \right) \times \max_{1 \le l \le K} \left(\sqrt{|S_i^l|} \right),$$

where $\tilde{b}_i$ is the virtual bid of buyer i, Since the buyer i is a winner, we have $\tilde{b}_i \ge cr(i)$ and thus $u_i \ge 0$. Buyer utility is always non-negative, which is not worse than staying outside the auction (i.e., utility is 0). Therefore, SMASHER-AP satisfies individual rationality.

Since SMASHER satisfies both incentive compatibility and individual rationality, SMASHER-AP is a strategy-proof mechanism. Our claim holds. □

5.4.2.5 Approximation Ratio

Theorem 5.5. *The approximation ratio of SMASHER is $O(n\sqrt{m})$, where n is the number of buyers, m is the number of channels.*

Proof. Let $(\mathbb{W}_{OPT}, \mathbb{S}_{OPT})$ be the optimal channel allocation, and $(\mathbb{W}_{APP}, \mathbb{S}_{APP})$ be the allocation achieved by SMASHER-AP. The social welfare of the optimal solution and SMASHER-AP is $\sum_{i \in \mathbb{W}_{OPT}} v_i$ and $\sum_{i \in \mathbb{W}_{APP}} v_i$, respectively.

For each $i \in \mathbb{W}_{APP}$, we define

$$\mathbb{W}_{OPT}^i = \left\{ j \in \mathbb{W}_{OPT} \middle| \frac{b_j}{\max\limits_{1 \leq l \leq K}\left(\sqrt{\left|S_j''\right|}\right)} \leq \frac{b_i}{\max\limits_{1 \leq l \leq K}\left(\sqrt{\left|S_i''\right|}\right)}, \right.$$

$$\left. \left(\mathbb{S}_{OPT}^j \cap \mathbb{S}_{APP}^i \cap \mathbb{VC} \neq \varnothing\right)\right\} \tag{5.55}$$

to represent the buyers in $\mathbb{W}_{OPT}$, whose bundles in $\mathbb{S}_{OPT}$ cannot be granted in SMASHER-AP because of the existence of i.

Since every $j \in \mathbb{W}_{OPT}^i$ appears after i in the ordered list $\mathbb{L}_1$, we have

$$v_j \leq \frac{v_i \times \max\limits_{1 \leq l \leq K}\left(\sqrt{\left|S_j''\right|}\right)}{\max\limits_{1 \leq l \leq K}\left(\sqrt{\left|S_i''\right|}\right)}. \tag{5.56}$$

Summing over all $j \in \mathbb{W}_{OPT}^i$, we can get

$$\sum_{j \in \mathbb{W}_{OPT}^i} v_j \leq \frac{v_i}{\max\limits_{1 \leq l \leq K}\left(\sqrt{\left|S_i''\right|}\right)} \sum_{j \in \mathbb{W}_{OPT}^i} \max\limits_{1 \leq l \leq K}\left(\sqrt{\left|S_j''\right|}\right). \tag{5.57}$$

Using the Cauchy-Schwarz inequality, we can bound

$$\sum_{j \in \mathbb{W}_{OPT}^i} \max\limits_{1 \leq l \leq K}\left(\sqrt{\left|S_j''\right|}\right) \leq \sqrt{\left|\mathbb{W}_{OPT}^i\right|} \sqrt{\sum_{j \in \mathbb{W}_{OPT}^i} \max\limits_{1 \leq l \leq K}\left(\left|S_j''\right|\right)}. \tag{5.58}$$

By integrating inequations (5.57) and (5.58), we get

$$\sum_{j \in \mathbb{W}_{OPT}^i} v_j \leq \frac{v_i \sqrt{\left|\mathbb{W}_{OPT}^i\right|} \sqrt{\sum_{j \in \mathbb{W}_{OPT}^i} \max\limits_{1 \leq l \leq K}\left(\left|S_j''\right|\right)}}{\max\limits_{1 \leq l \leq K}\left(\sqrt{\left|S_i''\right|}\right)}. \tag{5.59}$$

Since $(\mathbb{W}_{OPT}, \mathbb{S}_{OPT})$ is the optimal channel allocation, the channel bundles allocated to any pair of buyers $i, j \in \mathbb{W}_{OPT}$ cannot overlap on any virtual channel: $\mathbb{S}_{OPT}^i \cap \mathbb{S}_{OPT}^j \cap \mathbb{VC} = \varnothing$. Every bundle allocated to $j \in \mathbb{W}_{OPT}^i$ in the optimal allocation intersects with $\mathbb{S}_{APP}^i$ at least one virtual channel. Consequently, there are at most $\max\limits_{1 \leq l \leq K}\left(\left|S_i''\right|\right)$ buyers in $\mathbb{W}_{OPT}^i$

$$|\mathbb{W}^i_{OPT}| \leq \max_{1 \leq l \leq K} \left(\left| S_i'^l \right| \right)$$

$$\Rightarrow \sqrt{|\mathbb{W}^i_{OPT}|} \leq \max_{1 \leq l \leq K} \left(\sqrt{\left| S_i'^l \right|} \right). \tag{5.60}$$

Since $\max_{1 \leq l \leq K} \left(\left| S_j'^l \right| \right) \leq m(n-1) + 1 + m = mn + 1$, we have

$$\sum_{j \in \mathbb{W}^i_{OPT}} \max_{1 \leq l \leq K} \left(\left| S_j'^l \right| \right) \leq n(mn + 1). \tag{5.61}$$

By integrating inequations (5.59)–(5.61), we get

$$\sum_{j \in \mathbb{W}^i_{OPT}} v_j \leq \sqrt{n(mn + 1)} v_i. \tag{5.62}$$

Since $\mathbb{W}_{OPT} \subseteq \bigcup_{i \in \mathbb{W}_{APP}} \mathbb{W}^i_{OPT}$, we finally get

$$\sum_{i \in \mathbb{W}_{OPT}} v_i \leq \sqrt{n(mn + 1)} \sum_{i \in \mathbb{W}_{APP}} v_i. \tag{5.63}$$

Therefore, the approximation ratio of SMASHER is $O(n\sqrt{m})$. □

5.4.3 Evaluation Results

We implement SMASHER and compare its performance with TAHES [22] and CRWDP [17]. We also show the optimal results with tolerance 10^{-4}, denoted by IP-OPT, computed by solving the binary integer program in Sect. 5.4.1.2, as references of upper bound. Buyers are randomly distributed in a terrain area of $2,000 \times 2,000$ m. The number of buyers varies from 20 to 400 with increment of 20. The number of leasing channels can be one of the three values: 6, 12, and 24. The heterogeneous channels have different interference ranges, spanning from 250 to 450 m. We allow buyers to be equipped with different number of radios in our auctions, but limit the maximum size of channel bundle requested to 3. We assume that the buyers' channel bundle valuations are randomly distributed over $(0, 1]$. We consider the case of single-minded buyers (i.e., $K = 1$), and the case of multi-minded buyers who can submit up to 3 bundle requests (i.e., $K = 3$). All the results of performance are averaged over 200 runs.

Figure 5.12 shows the evaluation results when there are 12 channels and different number of buyers. We can see that SMASHER always outperforms the other two auction mechanisms, and its performance approaches the optimum, especially when $K = 1$. When the number of nodes is smaller than 60, TAHES cannot form

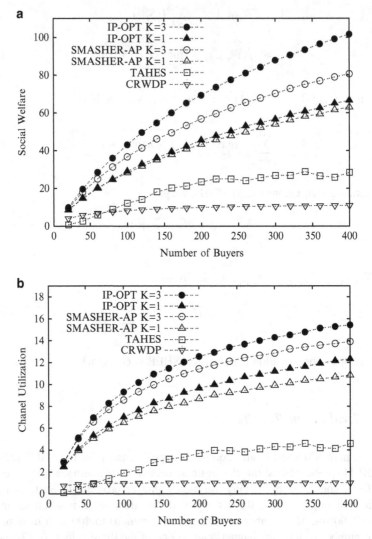

Fig. 5.12 Performance of SMASHER, TAHES, CRWDP and IP-OPT, when there are 12 channels. (**a**) Social welfare (**b**) Channel utilization

sufficient buyer groups with a larger number of bids, and thus does not perform well in this case. When the number of buyer is larger than 60, CRWDP's performance is not good because CRWDP does not consider channel spatial reusability (i.e., the channel utilization of CRWDP is equal to 1 in all cases). Figure 5.12 also shows that when the buyer number increases, the social welfare and channel utilization increase. SMASHER can allocate channels more efficiently among relatively large number of buyers, hence the social welfare and channel utilization increase.

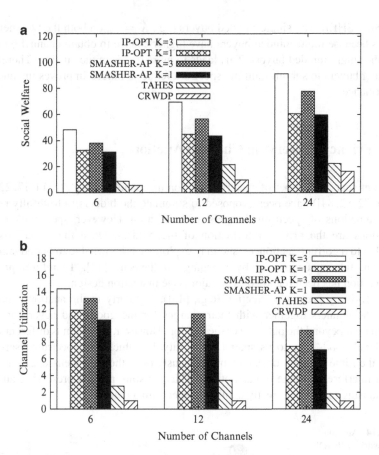

Fig. 5.13 Performance of SMASHER, TAHES, CRWDP and IP-OPT, when there are 200 buyers. (**a**) Social welfare (**b**) Channel utilization

Figure 5.13 shows the evaluation results when there are 200 buyers and the number of channels is 6, 12, and 24. Again, SMASHER-AP always achieves better performance than TAHES and CRWDP, whenever $K = 1$ or $K = 3$. Figure 5.13 also shows that when the number of channels increases, the social welfare increases and the channel utilization decreases. The reason is that larger supply of leasing channels leads to more trades in the auction, thus the social welfare increases when there exists a fixed number of buyers. The channel utilization decreases because buyers' radios can be allocated to more channels when the number of channels increases.

From Figs. 5.12 and 5.13, we can see that SMASHER sacrifices limited system performance to achieve economic robustness. Although IP-OPT achieves near optimal social welfare, we cannot apply it to channel redistribution problem, because IP-OPT does not has any guarantee on economic properties. We observe that SMASHER with multi-minded buyers (i.e., $K = 3$) always performs better than SMASHER with single-minded buyers (i.e., $K = 1$), on both the two metrics. This is because multi-minded buyers have higher chance to obtain channel bundles than the single-minded buyers. This leads to more trades in the auction. Therefore, allowing buyers to submit multiple spectrum requests indeed improves the auction performance.

5.5   Privacy Preserving Channel Auction

In recent years, a number of spectrum auction mechanisms (e.g., [3, 14, 17, 22, 61, 65, 71, 72, 82, 84]) have been proposed to stimulate the bidders to truthfully reveal their valuations of spectrum/channels in the auction. However, spectrum/channel valuations are the private information of the bidders. Once the valuations are revealed to a corrupt auctioneer, she may exploit such knowledge to her advantage, either in future auctions or by reneging on the sale [43]. Therefore, privacy preservation has been regarded as a major issue in auction design.

In a privacy preserving auction (e.g., [43]), any party in the auction can only know the winners together with their charges for the goods and never gain any information beyond the outcome of the auction. However, spectrum is different from traditional goods, due to its spatial reusability, by which two spectrum users can share the same wireless channel simultaneously once they are well-separated (i.e., out of interference range of each other). Thus, existing privacy preserving auction mechanisms cannot be directly applied to spectrum auctions.

Fig. 5.14 Auction framework of PRIDE

Auctioneer ⟺ Agent ⟺ Bidders

In this section, we consider the joint problem of designing both privacy preserving and strategy-proof auction mechanisms for spatially reusable radio spectrum. We present PRIDE, which is a *PRI*vacy preserving an*D* strat*E*gy-proof spectrum auction mechanism. As shown in Fig. 5.14, we introduce an agent in PRIDE, who can interact with both the auctioneer and the bidders. Bidders simultaneously submit their bids (encrypted by the method proposed in this paper) for channels via the agent to the auctioneer, such that no bidder can learn other participants' bids. The auctioneer decides the allocation of channels and the charges for the winners. The information stored at both the auctioneer and the agent is protected by cryptographic tools, such that neither of them can infer any sensitive information without the help of the other. As long as the agent and the auctioneer do not collude, PRIDE can guarantee both privacy preservation and strategy-proofness. Specifically,

PRIDE guarantees k-anonymous privacy preservation in a generic strategy-proof spectrum auction mechanism (e.g., [65, 82]).

In the field of privacy preservation, k-anonymity [54] is a commonly used criteria for evaluating privacy preserving schemes. A scheme provides k-anonymous protection when a person cannot be distinguished from at least $k - 1$ other individuals.

Definition 5.4 (k-**Anonymity** [54]). A privacy preserving scheme satisfies k-anonymity, if a participant cannot be identified by the sensitive information with probability higher than $1/k$.

5.5.1 A Generic Strategy-Proof Channel Auction

In this subsection, we present a generic strategy-proof channel auction mechanism, which is general enough to capture the essence of a category of strategy-proof channel auction mechanisms (e.g., [65, 82]). The generic spectrum auction presented here works in the case of single channel auction. In Sect. 5.5.4, we will show how to extend it to adapt to multi-channel bids.

In the generic spectrum auction, a channel can be leased to several bidders if they can transmit and receive signals simultaneously with an adequate signal to interference and noise ratio (SINR). We model the interference of the bidders by a conflict graph. Bidders are first divided into non-conflicting groups by any existing graph coloring algorithm (e.g., [64]) in a bid-independent way $\mathbb{G} = \{g_1, g_2, \ldots, g_q\}$, s.t., $g_j \cap g_l = \emptyset, \forall g_j, g_l \in \mathbb{G}, j \neq l$ and $\bigcup_{g_j \in \mathbb{G}} g_j = \mathbb{N}$.

A group bid σ_j for each group $g_j \in \mathbb{G}$ is calculated as

$$\sigma_j = |g_j| \min\{b_i | i \in g_j\}. \tag{5.64}$$

All bidder groups are ranked by their group bids in non-increasing order with bid-independent tie breaking:

$$\mathbb{G}' : g_1', g_2', \ldots, g_q', s.t., \sigma_1' \geq \sigma_2' \geq \ldots \geq \sigma_q'. \tag{5.65}$$

Bidders from the top $w = \min(m, q)$ groups are winners. Each winning group is charged with σ_{w+1}' (0, if σ_{w+1}' does not exist). The charge is shared evenly among the bidders in each winning group. Formally, a bidder i from a winning group g_j is charged with price

$$p_i = \begin{cases} \sigma_{w+1}'/|g_j|, & \text{if } q > m, \\ 0, & \text{otherwise.} \end{cases} \tag{5.66}$$

Algorithm 9 1-out-of-z oblivious transfer (OT_z^1)

1: **Initialization:**
2:　　**System parameters:** (g, h, G_t).
3:　　**Sender's input:** $s_1, s_2, \cdots, s_z \in G_t$.
4:　　**Receiver's choice:** $\alpha, 1 \leq \alpha \leq z$.
5: Receiver sends $y = g^r h^\alpha, r \in_R Z_t$.
6: Sender sends $c_i = (g^{k_i}, s_i(y/h^i)^{k_i}), k_i \in_R Z_t, 1 \leq i \leq z$.
7: By $c_\alpha = (d, f)$, receiver computes $s_\alpha = f/d^r$.

Essentially, the generic channel auction guarantees strategy-proofness, because the charge for a winner is independent of her bid.

Theorem 5.6. *The generic channel auction is a strategy-proof mechanism.*

5.5.2　Cryptographic Tools

We employ three cryptographic tools, including order preserving encryption, oblivious transfer, and secure multi-party computation.

5.5.2.1　Order Preserving Encryption

OPES [2] is a representative scheme to encrypt numeric data while preserving the order. It enables any comparison operation to be directly applied on the encrypted data.

Intuitively, we can protect the privacy of bidders in the auction by encrypting the bids in a way that preserves the order of bids and carrying out comparisons directly on the cipher text/value.

5.5.2.2　Oblivious Transfer

Oblivious transfer (OT) [47] describes a paradigm of secret exchange between two parties, a sender and a receiver.

The receiver can access one of the z secrets from the sender, without getting any information about the remaining $z - 1$ secrets, while the sender has no idea which of the z secrets was accessed. Algorithm 9 shows the pseudo-code of OT_z^1 proposed in [55], where t is a large prime, g and h are two generators of G_t, which is a cyclic group of order t, and Z_t is a finite additive group of t elements. As long as $\log_g h$ is not revealed, g and h can be used repeatedly. PRIDE employs an efficient 1-out-of-z oblivious transfer (OT_z^1) of integers [55].

5.5.2.3 Secure Multi-Party Computation

SMC, first proposed by Yao [74], has recently become appropriate for some realistic scenarios. We employ SMC in PRIDE to locate the lowest bid in each group. It enables a number of participants to carry out comparisons while preserving the privacy of their inputs.

5.5.3 Auction Design

In this section, we present the design of PRIDE.

5.5.3.1 Design Rational

PRIDE integrates cryptographic tools with the generic spectrum auction mechanism to achieve both strategy-proofness and privacy preservation. The main idea of PRIDE is to separate the information known by different parties in the auction, so that no party in the auction has enough knowledge to infer any sensitive information with confidence higher than $1/k$, while maintaining the functionality of the generic spectrum auction. We illustrate the design challenges and our idea in this subsection.

1. Information Separation

If there is a single central authority (auctioneer) carrying out the auction, it is inevitable that the sensitive information (i.e., each bidder's bid) is revealed to the auctioneer. To prevent this threat, we introduce a new entity, called agent. It is the agent's duty to tell the auctioneer the minimal amount of information necessary for deciding the winners and their charges. However, the information should not be fully accessed by the agent to prevent sensitive information leakage. So, we apply an end-to-end asymmetric encryption scheme between the auctioneer and the bidders, so that the agent cannot decrypt the bidding messages.

2. Bid Encryption

Since the auctioneer needs to find the lowest bid in each bidder group without knowing the exact values of bids from group members, we need a method to map the bids from the bidding space to another value space, while maintaining the comparison relation. We integrate the idea of order preserving encryption to enable such a mapping and prevent the auctioneer from learning the distribution of bids. We let the agent do the order preserving encryption before the auction. When bidding, the bidders contact the agent to get the mapped bids via oblivious transfer, which prevents the agent from knowing which bids are chosen. Later, the agent collects end-to-end encrypted bidding messages from bidders. Only the auctioneer can decrypt the bidding messages, extract mapped bids, and find the lowest mapped bid. The auctioneer can consult the agent to get the original value of the lowest mapped bid.

3. Outcome Verification

Different from traditional privacy preserving auctions, it is not easy for bidders to verify the correctness of auction outcome. We adopt the idea of SMC [74] to enable bidders from the same group to find the lowest bid, and thus verify the auction outcome.

5.5.3.2   Design Details

PRIDE works in four steps shown as follows.

Step 1: Initialization

Before running the spectrum auction, PRIDE sets up necessary system parameters. PRIDE defines a set of possible bid values as

$$\beta = \{\beta_1, \beta_2, \ldots, \beta_z\}, \tag{5.67}$$

in which $\beta_1 < \beta_2 < \ldots < \beta_z$, and requires that each bidder i's bid $b_i \in \beta$.

The agent maps each bid value $\beta_x \in \beta$ to γ_x, while maintaining the order, using the order preserving encryption scheme OPES.

$$\gamma_x = OPES(\beta_x), \text{ s.t., } \gamma_1 < \gamma_2 < \ldots < \gamma_z. \tag{5.68}$$

Here, $\gamma = \{\gamma_1, \gamma_2, \ldots, \gamma_z\}$ is a set of secrets of the agent. The agent also initializes the parameters of oblivious transfer by determining the large prime t and two generators of cyclic group G_t: (g, h).

PRIDE employs an asymmetric key encryption scheme. We suppose that the auctioneer holds a private key Key_{priv}, and the matching public key Key_{pub} is distributed to the bidders. PRIDE also employs a digital signature scheme, in which each bidder $i \in \mathbb{N}$ holds a signing key sk_i, and publishes the corresponding verification key pk_i.

Step 2: Bidding

Each bidder $i \in \mathbb{N}$ chooses a bid $b_i = \beta_x \in \beta$ according to her per channel valuation v_i, and then interacts with the agent through a 1-out-of-z oblivious transfer to receive $\hat{b}_i = \gamma_x$, which is the order-preserving-encrypted value of β_x.

- Bidder i randomly picks $r \in Z_t$, and sends $y = g^r h^x$ to the agent.
- The agent replies with $c = \{c_1, c_2, \ldots, c_z\}$, in which

$$c_l = \left(g^{k_l}, \gamma_l \left(y/h^l \right)^{k_l} \right), k_l \in_R Z_t, 1 \leq l \leq z. \tag{5.69}$$

Table 5.1 Information
published by the agent

Group ID	Bidder ID	Encrypted bid
1	$1_1, 1_2, \ldots, 1_{|g_1|}$	$e_{1,1}, e_{1,2}, \ldots, e_{1,|g_1|}$
2	$2_1, 2_2, \ldots, 2_{|g_2|}$	$e_{2,1}, e_{2,2}, \ldots, e_{2,|g_2|}$
$\vdots$	$\vdots$	$\vdots$
q	$q_1, q_2, \ldots, q_{|g_q|}$	$e_{q,1}, e_{q,2}, \ldots, e_{q,|g_q|}$

- The bidder picks $c_x = (d, f)$ from c, and computes

$$\hat{b}_i = \frac{f}{d^r} = \frac{\gamma_x (y/h^x)^{k_x}}{(g^{k_x})^r} = \frac{\gamma_x (g^r h^x / h^x)^{k_x}}{(g^{k_x})^r} = \gamma_x. \tag{5.70}$$

Upon receiving $\hat{b}_i$, bidder i randomly encrypts $\hat{b}_i$ using the auctioneer's public key Key_{pub}:

$$e_i = Encrypt\left(\hat{b}_i, Key_{pub}\right), \tag{5.71}$$

where $Encrypt()$ is the asymmetric encryption function. Bidder i then submits the following tuple as a bid to the agent

$$[i, e_i, Sign(e_i, sk_i)],$$

where $Sign()$ is the signing function.

For each tuple $[i, e_i, sign_i]$ received, the agent checks its validity. If

$$Verify(e_i, sign_i, pk_i) = True, \tag{5.72}$$

where $Verify()$ is the signature verification function, the tuple is accepted. Otherwise, it is discarded.

After collecting all the bids, the agent groups the bidders in a bid-independent way, as in the generic strategy-proof spectrum auction, and publishes the grouping result and encrypted bids, as shown in Table 5.1. To satisfy k-anonymity, we require that each bidder group must contain at least $k + 1$ bidders. In the table, bidder j_i is the ith member in group g_j, and $e_{j,1}, e_{j,2}, \ldots, e_{j,|g_j|}$ are encrypted bids from bidders in group g_j. Note that the order of $e_{j,i}$'s is irrelevant to the sequence of bidders in group g_j, which means that there is no one-to-one correspondence between $e_{j,i}$ and bidder j_i in any group.

Step 3: Opening

For each group $g_l \in \mathbb{G}$, the auctioneer decrypts the bids using her private key to get $\left\{\hat{b}_{l,1}, \hat{b}_{l,2}, \ldots, \hat{b}_{l,|g_l|}\right\}$:

$$\hat{b}_{l,i} = Decrypt\left(e_{l,i}, Key_{priv}\right), \forall i \in g_l, \tag{5.73}$$

where $Decrypt()$ is the asymmetric decryption function.

Since $\hat{b}_{l,i}$'s are computed by the order preserving encryption scheme, the lowest bid in group g_l must also be mapped to the smallest order-preserving-encrypted bid in g_l. Therefore, the auctioneer can locate the lowest bid $\hat{b}_l^{min}$ in group g_l by finding the smallest one in $\left\{\hat{b}_{l,1}, \hat{b}_{l,2}, \ldots, \hat{b}_{l,|g_l|}\right\}$:

$$\hat{b}_l^{min} = \min\left\{\hat{b}_{l,i} | 1 \le i \le |g_l|\right\}. \tag{5.74}$$

Then, the auctioneer resorts to the agent to fetch the original value b_l^{min} of $\hat{b}_l^{min}$:

$$b_l^{min} = OPES^{-1}(\hat{b}_l^{min}), \tag{5.75}$$

where $OPES^{-1}()$ is the reverse function of the order preserving encryption scheme. The auctioneer now can calculate the group bid of g_l:

$$\sigma_l = |g_l| b_l^{min}. \tag{5.76}$$

Similarly, the auctioneer calculates the group bids $\sigma_1, \sigma_2, \ldots, \sigma_q$ and sorts them in non-increasing order:

$$\sigma_1' \ge \sigma_2' \ge \ldots \ge \sigma_q'. \tag{5.77}$$

Same as the generic strategy-proof spectrum auction, winners $\mathbb{W}$ are the bidders from top $w = \min(m, q)$ groups:

$$\mathbb{W} = \bigcup_{j=1}^{w} g_j', \tag{5.78}$$

where g_j' is the group with jth highest group bid. In order to achieve strategy-proofness, each winning bidder group is charged with the group bid σ_{w+1}' of the $(w + 1)$th group (we set $\sigma_{w+1}' = 0$, if the $(w + 1)$th group does not exist.) The charge is shared evenly among all group members, hence each bidder i in winning group g_l is charged with

$$p_i = \sigma_{w+1}'/|g_l|. \tag{5.79}$$

Besides the set of winners $\mathbb{W}$ and their charges $(p_i)_{i \in \mathbb{W}}$, the auctioneer also announces σ_{w+1}' for public verification.

Step 4: Verification

This is an optional step. Any bidder group g_l, in which bidders doubt the outcome of the auction, can figure out the lowest bid $b_l^{min} = \min\{b_i | i \in g_l\}$ in the group by SMC [74] without disclosing their own inputs. Then the relation between $b_l^{min}|g_l|$ and σ'_{w+1} can be verified.

Figure 5.15 shows the message flow in PRIDE.

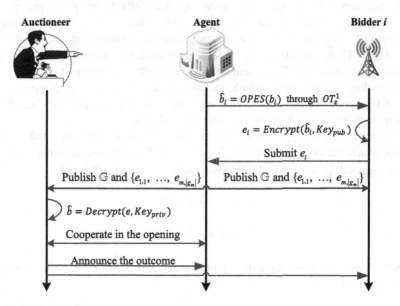

Fig. 5.15 Message flow

5.5.3.3 Illustrative Example

The following example may help to illustrate our mechanism. Figure 5.16 shows the interference range of seven bidders (A–G). They are competing for one channel. Assume that $\beta = \{1, 2, 3, 4, 5, 6, 7, 8, 9\}$ and the number beside each bidder

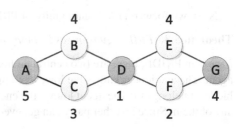

Fig. 5.16 Conflict graph

Table 5.2 Information
published by the agent

Group ID	Bidder ID	Encrypted bid
1	A, D, G	e_D, e_A, e_G
2	B, C, E, F	e_E, e_F, e_B, e_C

denotes her bid. For clarity and simplicity, we ignore the procedures of digital
signature/verification.

In the initialization step, the agent applies $OPES$ on β to get $\gamma =$
$\{3, 7, 10, 11, 15,\ 20, 23, 35, 90\}$. The seven bidders interact with the agent through
a 1-out-of-9 oblivious transfer to receive their order-preserving-encrypted bids
(i.e., $\hat{b}_A = 15, \hat{b}_B = 11, \dots, \hat{b}_G = 11$). Each bidder i encrypts her $\hat{b}_i$ with the
auctioneer's public key Key_{pub} and submits the result e_i to the agent.

According to the conflict graph, the bidders are split into two groups: $g_1 =$
$\{A, D, G\}$, $g_2 = \{B, C, E, F\}$. The agent publishes the grouping result and the
encrypted bids from each group, as shown in Table 5.2.

The auctioneer decrypts the encrypted bids and locates the lowest bid in each
group, which turns out to be $\hat{b}_1^{min} = 3, \hat{b}_2^{min} = 7$. Then she resorts to the agent for
the original values of $\hat{b}_1^{min}$ and $\hat{b}_2^{min}$, resulting in $b_1^{min} = 1, b_2^{min} = 2$.

$$\sigma_1 = 3 \times 1 = 3,$$

$$\sigma_2 = 4 \times 2 = 8,$$

thus $\sigma_2 > \sigma_1$. Therefore, g_2 is the winning group and B, C, E, F each is charged
with $\sigma_1/4 = 3/4$.

5.5.3.4   Analysis

We will show the strategy-proofness, k-anonymity, as well as some other attractive
properties of PRIDE.

The strategy-proofness of PRIDE is inherited from the generic strategy-proof
spectrum auction. Therefore, we omit the proof here and directly draw the following
conclusion, due to limitations of space.

Theorem 5.7. *PRIDE is a strategy-proof channel auction mechanism.*

Next, we focus on the k-anonymity of PRIDE.

Theorem 5.8. *PRIDE guarantees k-anonymity.*

Proof. In PRIDE, there are two central authorities, including the auctioneer and the
agent. The auctioneer knows the lowest bid in each group, but does not know which
bidder it belongs to. The agent knows the encrypted bids, but has no way to decrypt
any of them. Since no other party can get even more information than the auctioneer

or the agent, we focus on privacy protection against the auctioneer and the agent in this proof. We recall that each valid bidder group must contain at least $k+1$ bidders.

We distinguish the following two cases:

- **Case 1:** Bidder i belongs to a bidder group g_l that is satisfied with the outcome of the auction.

  On one hand, bidder i gets $\hat{b}_i = \gamma_x$ through a 1-out-of-z oblivious transfer from the agent, who is unaware of which γ_x has been accessed by the bidder. Bidder i then sends the encrypted bid e_i to the agent, who cannot decrypt e_i without knowing the private key of the asymmetric encryption scheme. Although the agent may know the lowest bid in group g_l later when the auctioneer consults her, she still cannot infer its bidder. So, the agent can not identify the bidder of the lowest bid in group g_l from at least $k+1$ bidders.

  On the other hand, although the auctioneer can decrypt an anonymous ciphertext e to get $\hat{b}$, she can only reversely map the lowest $\hat{b}_{min}$ to the original bid b_{min} for each group, resorting to the agent. However, the auctioneer still cannot infer the bidder, to which b_{min} belongs out of at least $k+1$ members in group g_l.

  So, neither the agent, nor the auctioneer, can identify any bidder's bid with probability higher than $1/(k+1)$.
- **Case 2:** Bidder i belongs to a bidder group g_l, which wants to verify the auction outcome. This case only diverges from the previous one in the public verification step. Therefore, we focus on the verification step here.

  Since secure multi-party computation is applied to find the lowest b_l^{min} in group g_l, any group member cannot identify the owner of b_l^{min} from the rest k bidders.

Therefore, we can conclude that PRIDE guarantees k-anonymity. □

Besides strategy-proofness and k-anonymity, PRIDE also achieves the following nice properties.

- *Public Verifiability*: It enables bidder groups to verify the outcome of the auction in public verification step.
- *Non-Repudiation*: No bidder can deny her bid after the auction, since her signature is required to be verified when the bidder submits her bid to the agent.
- *Low Communication Overhead*: When z is constant, the communication overhead induced by PRIDE is $O(n)$, where n is the number of bidders.
- *Low Computation Overhead*: The cryptographic tools adopted by PRIDE are light weighted schemes, which only induce a small amount of computation overhead. Our evaluation results show that the computation overhead of PRIDE is rather low.

5.5.4 Extension to Multi-Channel Bids

In the previous section, we present a strategy-proof and privacy preserving auction mechanism, in which each bidder bids for a single channel. In this section, we extend PRIDE to adapt to the scenario in which a bidder can bid for multiple channels. Similarly, our extension achieves both strategy-proofness and k-anonymity.

We now allow each bidder $i \in \mathbb{N}$ to demand d_i channels. Let $\mathbf{d} = (d_1, d_2, \ldots, d_n)$ denote the demand profile of bidders.

We assume that each bidder has an identical valuation on different channels. In the auction, each bidder i submits not only her encrypted bid per channel b_i, but also the number of channels demanded d_i. We also assume that the bidders do not cheat the demands for two reasons. On one hand, the auction only allocates the channels to the bidders up to their demands. A bidder's demand definitely cannot be contented if she lowers the demand. On the other hand, over demanding may result in winning more than enough channels. Although the bidder has no valuation on the extra channels, she still needs to pay for them.

To extend PRIDE to adapt to multi-channel bids, we introduce *virtual group*, and update bidding and opening steps of PRIDE. Note that the basic version of PRIDE presented in Sect. 5.5.3 is a special case of the extended PRIDE.

5.5.4.1 Virtual Group

In the extended PRIDE, the bidders from the same group may demand different numbers of channels. To represent the various demands in a bidder group, we introduce the concept of *virtual group*.

Given a bidder group $g_l \subseteq \mathbb{N}$, let $\hat{d}_l$ be the maximum channel demand in group g_l:

$$\hat{d}_l = \max\{d_i | i \in g_l\}. \tag{5.80}$$

A virtual group $\tilde{g}_l^j \subseteq g_l$ is the set of bidders, who demand at least j channels in bidder group g_l:

$$\tilde{g}_l^j = \{i | i \in g_l \land d_i \geq j\}, 1 \leq j \leq \hat{d}_l. \tag{5.81}$$

Algorithm 10 shows the pseudo-code of virtual group generation. We find the maximum channel demand $\hat{d}_l$ in group g_l (lines 2–4), and iteratively pick the bidders demanding at least j channels to form virtual group $\tilde{g}_l^j$, which is added into the set G_l of virtual groups generated from group g_l (lines 5–8). Figure 5.17 shows our idea of virtual group generation. Each number in a circle denotes the demand of a bidder.

Algorithm 10 Virtual group generation—vgrouping(g_l)

Input: Bidder group g_l, demand profile **d**.
Output: Set of virtual groups G_l.
1: $G_l \leftarrow \emptyset; \hat{d}_l \leftarrow 0$.
2: **for all** $i \in g_l$ **do**
3:      $\hat{d}_l \leftarrow \max(\hat{d}_l, d_i)$.
4: **end for**
5: **for** $j \leftarrow 1, \ldots, \hat{d}_l$ **do**
6:      $\tilde{g}_l^j \leftarrow \{i \mid i \in g_l \wedge d_i \geq j\}$.
7:      $G_l \leftarrow G_l \cup \{\tilde{g}_l^j\}$.
8: **end for**
9: Return G_l.

Fig. 5.17 A toy example

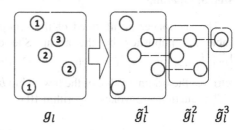

$$g_l \qquad\qquad \tilde{g}_l^1 \quad \tilde{g}_l^2 \quad \tilde{g}_l^3$$

In the extended PRIDE, an original bidder group g_l is replaced by $\hat{d}_l$ virtual groups. The group bid $\tilde{\sigma}_l^j$ of virtual group $\tilde{g}_l^j$ is defined as

$$\tilde{\sigma}_l^j = \left| \tilde{g}_l^j \right| \min\{b_i \mid i \in g_l\}. \tag{5.82}$$

Note that in order to guarantee k-anonymity, the lowest bid in group g_l, instead of virtual group $\tilde{g}_l^j$, is used to calculate the group bids of virtual groups.

5.5.4.2 Extension Details

The procedures of initialization and verification are the same as those in the basic PRIDE. Due to limitations of space, we focus on the differences in the steps of bidding and opening.

Step 1: Initialization

Please refer to Sect. 5.5.3.2 for details.

Step 2: Bidding

In order to include the information of channel demands, the tuple submitted by bidder i to the agent must have one more element d_i:

$$[i, e_i, d_i, Sign(e_i \| d_i, sk_i)],$$

where $\|$ is the concatenation operation.

Table 5.3 Information published by the agent

Group ID	Bidder ID & Demand	Encrypted bid
1	$[1_1, d_{1_1}], \ldots, [1_{|g_1|}, d_{|g_1|}]$	$e_{1,1}, \ldots, e_{1,|g_1|}$
2	$[2_1, d_{2_1}], \ldots, [2_{|g_2|}, d_{|g_2|}]$	$e_{2,1}, \ldots, e_{2,|g_2|}$
$\vdots$	$\vdots$	$\vdots$
q	$[q_1, d_{q_1}], \ldots, [q_{|g_q|}, d_{|g_q|}]$	$e_{q,1}, \ldots, e_{q,|g_q|}$

The agent collects the bidding messages, verifies the validity, and publishes the grouping results and encrypted bids. This time, beside each bidder's ID, there is a corresponding channel demand, as shown in Table 5.3.

Step 3: Opening

The auctioneer is informed of the grouping results and encrypted bids from Table 5.3 published by the agent. She decrypts the encrypted bids to get $\{\hat{b}_{l,1}, \hat{b}_{l,2}, \ldots, \hat{b}_{l,|g_l|}\}$ for each $g_l \in \mathbb{G}$. Resorting to the agent, the auctioneer retrieves the original value of the lowest bid b_l^{min} of each $g_l \in \mathbb{G}$.

The auctioneer invokes Algorithm 10 to form virtual groups:

$$\tilde{\mathbb{G}} = \bigcup_{g_l \in \mathbb{G}} G_l. \tag{5.83}$$

For each virtual group $\tilde{g}_l^j \in \tilde{\mathbb{G}}$, the auctioneer calculates the virtual group bid:

$$\tilde{\sigma}_l^j = \left| \tilde{g}_l^j \right| b_l^{min}. \tag{5.84}$$

Next, the auctioneer sorts all the virtual groups according to their group bids in non-increasing order:

$$\tilde{\sigma}_1'' \geq \tilde{\sigma}_2'' \geq \ldots \geq \tilde{\sigma}_{\sum_{g_l \in \mathbb{G}} \hat{d}_l}''. \tag{5.85}$$

Auction winners $\mathbb{W}'$ are the bidders in the top $w' = \min(m, \sum_{g_l \in \mathbb{G}} \hat{d}_l)$ virtual groups:

$$\mathbb{W}' = \bigcup_{j=1}^{w'} g_j'', \tag{5.86}$$

Algorithm 11 Charging algorithm—charging(i)

Input: Set of virtual groups $\tilde{\mathbb{G}}$ and corresponding virtual group bids $\left(\tilde{\sigma}_l^j\right)_{\tilde{g}_l^j \in \tilde{\mathbb{G}}}$, winner $i \in g_l$.

Output: Charge p_i.

1: $\tilde{\mathbb{G}}' \leftarrow \tilde{\mathbb{G}} \setminus \left\{\tilde{g}_l^j | 1 \le j \le \hat{d}_l\right\}$.

2: Sort the virtual groups in $\tilde{\mathbb{G}}'$ by virtual group bid in non-increasing order $\sigma_1^\Delta \ge \sigma_2^\Delta \ge \dots \ge \sigma_{\sum_{g_k \in G \wedge i \notin g_k} \hat{d}_k}^\Delta$.

3: $p_i \leftarrow 0$.

4: **for** $h \leftarrow 1, \dots, a_i$ **do**

5:    $t \leftarrow \min\left(m - h + 1, \sum_{g_k \in G \wedge i \notin g_k} \hat{d}_k\right)$.

6:    **if** $t = m - h + 1$ **then**

7:       $p_i \leftarrow p_i + \sigma_t^\Delta / |\tilde{g}_l^h|$.

8:    **end if**

9: **end for**

10: Return p_i.

---

where g_j'' is the jth highest bid virtual group. The number of channels each bidder $i \in \mathbb{W}'$ wins is

$$a_i = \sum_{1 \le j \le w' \wedge i \in g_j''} 1. \tag{5.87}$$

Since a bidder may be in multiple virtual groups, the previous method of charging can no longer be applied. We present a new charging method as shown in Algorithm 11.

In Algorithm 11, we remove all the virtual groups generated from the bidder group, to which the winning bidder i belongs, and sort the rest virtual groups by virtual group bid in non-increasing order (lines 1–2). Then, for each channel h won by bidder i, we locate the virtual group in the sorted list, after which wins a channel, bidder i cannot win channel h. If such a virtual group does not exist, then channel h is free of charge for bidder i. Otherwise, the located virtual group's bid is used to calculate the charge for bidder i on channel h. The charge on channel h is set to $\sigma_t^\Delta / |\tilde{g}_l^h|$. The total charge for bidder i is the sum of charges on all the channels won (lines 3–9).

Finally, the auctioneer releases the set of winners $\mathbb{W}'$, the channel allocation profile **a**, and the charging profile **p**.

Step 4: Verification

Please refer to Sect. 5.5.3.2 for details.

Again, we show that PRIDE satisfies both strategy-proofness and k-anonymity, in the case of multi-channel bids.

Theorem 5.9. *PRIDE is a strategy-proof channel auction mechanism, despite of multi-channel bids.*

Proof. We consider an arbitrary bidder $i \in g_l$ in the auction. Her utility is

$$u_i = v_i a_i - p_i = v_i a_i - \sum_{h=1}^{a_i} p_i^h, \qquad (5.88)$$

where

$$p_i^h = \begin{cases} \sigma_{c-h+1}^{\Delta}/|\tilde{g}_l^h|, & \text{if } \sum_{g_k \in G \wedge i \notin g_k} \hat{d}_k \geq m - h + 1, \\ 0, & \text{otherwise.} \end{cases} \qquad (5.89)$$

Since p_i^h's are independent of the bidder i's bid b_i, the utility is a function on the number of allocated channels a_i.

Suppose a_i is the number of channels won by bidder i, when she bids truthfully, i.e., $b_i = v_i$. We then distinguish two cases:

- The bidder i wins more channels (i.e., $a_i' > a_i$) by bidding another value $b_i' \neq b_i$. This happens only when the bidder i holds the smallest bid in group g_l when bidding truthfully, and wins more channels by raising her bid (i.e., $b_i' > b_i$) to increase the virtual groups' bids. Let $h(a_i < h \leq a_i')$ be the hth additional channel won by the bidder i. Then $p_i^h > 0$, because otherwise the bidder would win this channel, when bidding truthfully. The utility got on this channel is

$$\begin{aligned}
u_i^h &= v_i - p_i^h \\
&= v_i - \sigma_{c-h+1}^{\Delta}/|\tilde{g}_l^h| \\
&= v_i - b_l'^{min}|\tilde{g}_l^h|/|\tilde{g}_l^h| \\
&= v_i - b_l'^{min} \\
&\leq v_i - b_i \\
&= 0.
\end{aligned}$$

Therefore, getting any more channel does not increase the bidder i's utility.
- The bidder i wins less channels (i.e., $a_i' < a_i$) by bidding another value $b_i' \neq b_i$. Since the charging algorithm guarantees that

$$p_i^h \leq b_i, \forall 1 \leq h \leq a_i,$$

the utility got on the hth channel is always non-negative

$$u_i^h = v_i - p_i^h \geq v_i - b_i = 0.$$

Therefore, losing any channel cannot benefit bidder i.

Consequently, bidding truthfully is every bidder's dominant strategy, and thus PRIDE satisfies incentive-compatibility.

Furthermore, since any bidder who loses in the auction is free of charge, and also since any winner is charged on each channel with price not exceeding her bid, PRIDE also satisfies individual-rationality.

Therefore, we can conclude that PRIDE is a strategy-proof spectrum auction mechanism, despite of multi-channel bids. □

Since PRIDE does not reveal any more information to any party, in the case of multi-channel bids, we have the following theorem.

Theorem 5.10. *PRIDE guarantees k-anonymity, despite of multi-channel bids.*

Besides strategy-proofness and k-anonymity, PRIDE for multi-channel bids also has good properties, including public verifiability, non-repudiation, low communication and computation overhead. Due to limitations of space, we do not illustrate the details again.

5.5.5 Evaluation Results

We have implemented PRIDE and evaluated its performance on the efficiency of spectrum auction and overheads introduced.

In the evaluations, we vary the number of bidders from 50 to 500, the number of channels from 5 to 50, and the terrain area from 500×500 m to $2,000 \times 2,000$ m. In each set of evaluations, we vary a factor among bidder number, channel number, and terrain area, and fix the other two. The default value for bidder number, channel number, and terrain area, is 200, 20, and $2,000 \times 2,000$ m, respectively. The bidders are randomly distributed in the terrain area, and the interference range is set to 425 m. In the case of multi-channel demand, we randomly generate the demand of each bidder from $\{1, 2, 3, 4, 5\}$.

5.5.5.1 Results on Channel Utilization

Figure 5.18 shows the evaluation results of PRIDE on channel utilization, when bidders can bid for single channel (PRIDE-SINGLE) and multiple channels (PRIDE-MULTIPLE).

Figure 5.18a shows the channel utilizations achieved by PRIDE, when we fix the number of channels and terrain area, and vary the number of bidders. Here we observe that, when the number of bidders is less than 200, the channel utilization of PRIDE-SINGLE is lower than that of PRIDE-MULTIPLE. This is because the channels are over supplied. When we allow the bidders to demand multiple channels, the channels can be better exploited. However, with growth of the number of bidders, especially when the number of bidders is larger than or equal to 200,

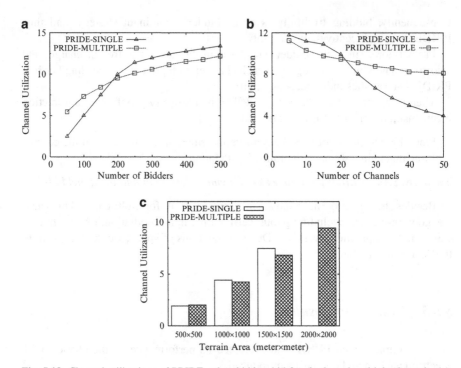

Fig. 5.18 Channel utilizations of PRIDE, when bidders bid for single and multiple channels. (**a**) Effect of number of bidders. (**b**) Effect of number of channels. (**c**) Effect of size of terrain area

the channels supplied become more and more scarce compared with the number of bidders, and the competition among the bidders become more and more intense. The introduction of virtual group makes the average (virtual) group size smaller than the single-channel bid case, and thus results in a lower channel utilization.

Figure 5.18b shows the channel utilizations achieved by PRIDE, when varying the number of channels and fixing the other two factors. When the number of channels is no more than 20, PRIDE-MULTIPLE has a lower channel utilization than PRIDE-SINGLE, due to the smaller average (virtual) group size. However, with more than 20 channels supplied, PRIDE-MULTIPLE has a higher channel utilization than PRIDE-SINGLE, due to higher demands from the bidders.

Figure 5.18c shows the case, in which we vary the size of terrain area and fix the other two factors. When the terrain area is 500×500 m or $1,000 \times 1,000$ m, most of the (virtual) groups contain only 1 or 2 bidders, thus the difference between PRIDE-SINGLE and PRIDE-MULTIPLE is very small. However, with the increment of terrain area, the difference between PRIDE-SINGLE and PRIDE-MULTIPLE on average size of (virtual) groups becomes larger and larger, resulting in the channel utilization of PRIDE-MULTIPLE lower than that of PRIDE-SINGLE.

5.5.5.2 Overhead

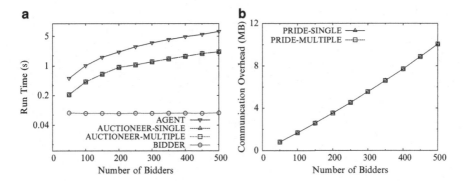

Fig. 5.19 Computation and communication overheads induced by PRIDE. (**a**) Computation overhead. (**b**) Communication overhead

We implement PRIDE using JavaSE-1.7 with packages java.security and javax.crypto, and use RSA with modulus of $1,024$ bits to do encryption/decryption and digital signature/verfication. Bidders can choose one out of $1,000$ predefined bids in the auction, and get 128 bits of order-preserving-encrypted value through oblivious transfer with the agent. The running environment is *Intel(R) Core(TM) i7* 2.67 GHz and Windows 7.

Figure 5.19a shows the computation overhead of the agent, the auctioneer, and each bidder, as a function of the number of bidders. We can see that the computation overhead is mainly on the agent, because the agent is responsible for oblivious transfer and bidder grouping. The computation overhead of agent is 0.515 s for 50 bidder, and 6.520 s for 500 bidders. The auctioneer has a lower computation overhead than the agent. The computation overhead of each bidder is very small.

Figure 5.19b shows the overall communication overhead induced by PRIDE. The communication overhead induced is mainly from the oblivious transfer. In the oblivious transfer, the agent needs to transfer 128 bits for each of the $1,000$ possible bids to every bidder.

Observing the computation and communication overheads shown above, we can conclude that the overheads induced by PRIDE is small enough to be applied to wireless devices.

Chapter 6
Summary and Open Problems

In this book, we have presented a systematic study on the problem of game-theoretic spectrum redistribution. We have studied the problem from a game-theoretic perspective, in which the nodes in the wireless network are strategic players with self-interests and have to compete with each other to get the limited communication spectrum.

In Chap. 4, we have focused on the problem of game-theoretic channel allocation in clique, where all the nodes are in the same collision domain, and have presented two complementary approaches for pre-partitioned (Sect. 4.1) and adjustable (Sect. 4.2) channel allocation, respectively. Both of the two approaches can achieve globally optimal network throughput, and guarantee the system converge to a (strongly) dominant strategy equilibrium in a single step.

In Chap. 5, we have considered the problem of game-theoretic channel allocation in large geographical area, in which the radio spectrum can be spatially reused, and have adopted the concept of auctions to perform highly efficient channel allocation. Four strategy-proof channel auction mechanisms have been presented, including SMALL (Sect. 5.2), SPECIAL (Sect. 5.3), SMASHER (Sect. 5.4), and PRIDE (Sect. 5.5). SMALL is a graph coloring-based sealed-bid reserve auction mechanism, which accepts uniform bids for a single or multiple channels. SPECIAL is a combinatorial channel auction mechanism accepting flexible bids for different numbers of contiguous channels. SMASHER is a combinatorial auction mechanism for heterogeneous channel redistribution, achieving approximately efficient social welfare. Finally, PRIDE is a privacy preserving and strategy-proof channel auction mechanism.

Although a large number of works on game-theoretic spectrum redistribution have been proposed, there still exist three major open problems.

1. Distributed Auction: All of the existing spectrum auction mechanisms rely on a trusted authority to serve as the auctioneer. However, it is not easy to implement the trusted authority in distributed wirelessnetworks. An interesting alternative is

F. Wu, *Game Theoretic Approaches for Spectrum Redistribution*, SpringerBriefs in Electrical and Computer Engineering, DOI 10.1007/978-1-4939-0500-3_6, © The Author(s) 2014

to design fully distributed spectrum auction mechanisms, which should be able to deal with strategic players' both computation manipulation and communication manipulation.

2. Revenue Maximization: In contrast to spectrum buyers' objectives, a spectrum seller's interest is to maximize her revenue for selling/leasing the spectrum. However, optimizing social welfare does not necessarily maximize the seller's revenue. Existing revenue maximizing works on spectrum auction either cannot guarantee strategy-proofness, or rely on a prior knowledge of distribution of the buyers' valuations, which make the auction mechanisms not practical.

3. Privacy Preservation: As we have mentioned, spectrum/channel valuations are private information of the bidders, and should be protected from both the auctioneer and the other competing bidders. Our preliminary privacy preserving spectrum auction mechanism only allow the bidders to choose bids from a small set of predefined values, in order to avoid high computation and communication overhead. Adopting homomorphic cryptographic tools can be a promising way to strengthen the existing work. Furthermore, bidders's geographic locations are also private information, and should be protected in the spectrum auction.

References

1. Adya, A., Bahl, P., Padhye, J., Wolman, A., Zhou, L.: A multi-radio unification protocol for ieee 802.11 wireless networks. In: Proceedings of the 1st International Conference on Broadband Networks (BroadNets), pp. 344–354 (2004)
2. Agrawal, R., Kiernan, J., Srikant, R., Xu, Y.: Order preserving encryption for numeric data. In: Proceedings of the 2004 ACM International Conference on Management of Data (SIGMOD), pp. 563–574, Paris, France (2004)
3. Al-Ayyoub, M., Gupta, H.: Truthful spectrum auctions with approximate revenue. In: Proceedings of 30th Annual IEEE Conference on Computer Communications (INFOCOM), Shanghai, China (2011)
4. Alicherry, M., Bhatia, R., Li, L.: Joint channel assignment and routing for throughput optimization in multi-radio wireless mesh networks. In: Proceedings of the 11th International Conference on Mobile Computing and Networking (MobiCom), Cologne, Germany (2005)
5. Ben Salem, N., Buttyan, L., Hubaux, J.P., Jakobsson, M.: A charging and rewarding scheme for packet forwarding in multi-hop cellular networks. In: Proceedings of the 4th ACM Symposium on Mobile Ad Hoc Networking and Computing (MobiHoc), Annapolis, MD (2003)
6. Bianchi, G.: Performance analysis of the IEEE 802.11 distributed coordination function. IEEE J. Sel. Areas Commun. **18**(3), 535–547 (2000)
7. Čagalj, M., Ganeriwal, S., Aad, I., Hubaux, J.P.: On selfish behavior in CSMA/CA networks. In: Proceedings of 24th Annual IEEE Conference on Computer Communications (INFOCOM), Miami, FL (2005)
8. Chandra, R., Mahajan, R., Moscibroda, T., Raghavendra, R., Bahl, P.: A case for adapting channel width in wireless networks. In: Proceedings of ACM SIGCOMM 2008 Conference on Applications, Technologies, Architectures, and Protocols for Computer Communications, Seattle, WA (2008)
9. Chen, X., Huang, J.: Game theoretic analysis of distributed spectrum sharing with database. In: Proceedings of the 33rd International Conference on Distributed Computing Systems (ICDCS), Philadelphia, USA (2012)
10. Chen, X., Huang, J.: Spatial spectrum access game: Nash equilibria and distributed learning. In: Proceedings of the 13th ACM Symposium on Mobile Ad Hoc Networking and Computing (MobiHoc), Hilton Head, USA (2012)
11. Chen, Y., Duan, L., Huang, J., Zhang, Q.: Balance of revenue and social welfare in FCC's spectrum allocation. In: Proceedings of 32nd IEEE International Conference on Computer Communications (INFOCOM), Turin, Italy (2013)
12. Clarke, E.H.: Multipart pricing of public goods. Public Choice **11**, 17–33 (1971)
13. Cox, D.C., Reudink, D.O.: Dynamic channel assignment in high capacity mobile communication system. Bell Syst. Tech. J. **50**(6), 1833–1857 (1971)

F. Wu, *Game Theoretic Approaches for Spectrum Redistribution*, SpringerBriefs
in Electrical and Computer Engineering, DOI 10.1007/978-1-4939-0500-3,
© The Author(s) 2014

14. Deek, L., Zhou, X., Almeroth, K., Zheng, H.: To preempt or not: tackling bid and time-based cheating in online spectrum auctions. In: Proceedings of 30th Annual IEEE Conference on Computer Communications (INFOCOM), Shanghai, China (2011)
15. Ding, L., Melodia, T., Batalama, S., Matyjas, J.: Distributed routing, relay selection, and spectrum allocation in cognitive and cooperative ad hoc networks. In: Proceedings of the 7th Annual IEEE Communications Society Conference on Sensor, Mesh and Ad Hoc Communications and Networks (SECON), Boston, USA (2010)
16. Dobzinski, S.: An impossibility result for truthful combinatorial auctions with submodular valuations. In: Proceedings of the 43rd Annual Symposium on Theory of Computing (STOC), San Jose, CA (2011)
17. Dong, M., Sun, G., Wang, X., Zhang, Q.: Combinatorial auction with time-frequency flexibility in cognitive radio networks. In: Proceedings of 31st Annual IEEE International Conference on Computer Communications (INFOCOM), Orlando, FL (2012)
18. Eidenbenz, S., Resta, G., Santi, P.: Commit: a sender-centric truthful and energy-efficient routing protocol for ad hoc networks with selfish nodes. In: Proceedings of the 19th International Parallel and Distributed Processing Symposium (IPDPS), Denver, CO (2005)
19. Federal Communications Commission (FCC): http://www.fcc.gov/
20. Félegyházi, M.: Non-cooperative behavior in wireless networks. Ph.D. thesis, EPFL, Switzerland (2007)
21. Félegyházi, M., Čagalj, M., Bidokhti, S.S., Hubaux, J.P.: Non-cooperative multi-radio channel allocation in wireless networks. In: Proceedings of 26th Annual IEEE Conference on Computer Communications (INFOCOM), Anchorage, AK (2007)
22. Feng, X., Chen, Y., Zhang, J., Zhang, Q., Li, B.: TAHES: Truthful double auction for heterogeneous spectrums. In: Proceedings of 31st Annual IEEE International Conference on Computer Communications (INFOCOM), Orlando, FL (2012)
23. Fudenberg, D., Tirole, J.: Game Theory. MIT Press, Cambridge (1991)
24. Garey, M.R., Johnson, D.S.: Computers and Intractability: A Guide to the Theory of NP-Completeness. W. H. Freeman & Co., New York (1990)
25. Gibbons, R.: A Primer in Game Theory. Prentice Hall, Englewood Cliffs (1992)
26. Gopinathan, A., Li, Z.: A prior-free revenue maximizing auction for secondary spectrum access. In: Proceedings of 30th Annual IEEE Conference on Computer Communications (INFOCOM), Shanghai, China (2011)
27. Gopinathan, A., Li, Z.: Strategyproof auctions for balancing social welfare and fairness in secondary spectrum markets. In: Proceedings of 30th Annual IEEE Conference on Computer Communications (INFOCOM), Shanghai, China (2011)
28. Groves, T.: Incentives in teams. Econometrica: J. Econom. Soc. **41**, 617–631 (1973)
29. Halldórsson, M.M., Halpern, J.Y., Li, L.E., Mirrokni, V.S.: On spectrum sharing games. In: Proceedings of the 23rd Annual ACM SIGACT-SIGOPS Symposium on Principles of Distributed Computing (PODC), St. John's, Canada (2004)
30. Han, B., Kumar, V., Marathe, M., Parthasarathy, S., Srinivasan, A.: Distributed strategies for channel allocation and scheduling in software-defined radio networks. In: Proceedings of 28th Annual IEEE Conference on Computer Communications (INFOCOM), Rio de Janeiro, Brazil (2009)
31. Hou, Y.T., Shi, Y., Sherali, H.D.: Optimal spectrum sharing for multi-hop software defined radio networks. In: Proceedings of 26th Annual IEEE Conference on Computer Communications (INFOCOM), Anchorage, AK (2007)
32. Huang, J., Berry, R.A., Honig, M.L.: Auction-based spectrum sharing. Mobile Networks Appl. **11**(3), 405–418 (2006)
33. Jia, J., Zhang, Q., Zhang, Q., Liu, M.: Revenue generation for truthful spectrum auction in dynamic spectrum access. In: Proceedings of the 10th ACM Symposium on Mobile Ad Hoc Networking and Computing (MobiHoc), New Orleans, LA (2009)
34. Kasbekar, G.S., Sarkar, S.: Spectrum pricing games with bandwidth uncertainty and spatial reuse in cognitive radio networks. In: Proceedings of the 11th ACM Symposium on Mobile Ad Hoc Networking and Computing (MobiHoc), Chicago, IL (2010)

35. Kasbekar, G.S., Sarkar, S.: Spectrum pricing games with spatial reuse in cognitive radio networks. IEEE J. Sel. Areas Commun. **30**(1), 153–164 (2012)
36. Katzela, I., Naghshineh, M.: Channel assignment schemes for cellular mobile telecommunications: a comprehensive survey. IEEE Personal Commun. **3**(3), 10–31 (1996)
37. Kodialam, M., Nandagopal, T.: Characterizing the capacity region in multi-radio multi-channel wireless mesh networks. In: Proceedings of the 11th International Conference on Mobile Computing and Networking (MobiCom), Cologne, Germany (2005)
38. Kyasanur, P., Vaidya, N.: A routing protocol for utilizing multiple channels in multi-hop wireless networks with a single transceiver. In: Proceedings of the 2nd International Conference on Quality of Service in Heterogeneous Wired/Wireless Networks (QShine), Orlando, FL (2005)
39. Lehmann, D., Oćallaghan, L.I., Shoham, Y.: Truth revelation in approximately efficient combinatorial auctions. J. ACM **49**(5), 577–602 (2002)
40. Mas-Colell, A., Whinston, M.D., Green, J.R.: Microeconomic Theory. Oxford University Press, Oxford (1995)
41. Mishra, A., Banerjee, S., Arbaugh, W.: Weighted coloring based channel assignment for WLANs. ACM SIGMOBILE Mobile Comput. Commun. Rev. (MC2R) **9**(3), 19–31 (2005)
42. Mishra, A., Brik, V., Banerjee, S., Srinivasan, A., Arbaugh, W.: A client-driven approach for channel management in wireless LAN. In: Proceedings of 25th Annual IEEE Conference on Computer Communications (INFOCOM), Barcelona, Spain (2006)
43. Naor, M., Pinkas, B., Sumner, R.: Privacy preserving auctions and mechanism design. In: Proceedings of the ACM Symposium on Electronic Commerce (EC) (1999)
44. Nash, J.: Equilibrium points in n-person games. In: Proceedings of the National Academy of Sciences, vol. 36, pp. 48–49 (1950)
45. von Neumann, J., Morgenstern, O.: Theory of Games and Economic Behavior. Princeton University Press, Princeton (1947)
46. Osborne, M.J., Rubenstein, A.: A Course in Game Theory. MIT Press, Cambridge (1994)
47. Rabin, M.O.: How to exchange secrets with oblivious transfer. Technical Report, Aiken Computation Lab, Harvard University (1981)
48. Radio Administration Bureau (RAB): http://wgj.miit.gov.cn/
49. Raman, B.: Channel allocation in 802.11-based mesh networks. In: Proceedings of 25th Annual IEEE Conference on Computer Communications (INFOCOM), Barcelona, Spain (2006)
50. Raniwala, A., Gopalan, K., cker Chiueh, T.: Centralized channel assignment and routing algorithms for multi-channel wireless mesh networks. ACM SIGMOBILE Mobile Comput. Commun. Rev. (MC2R) **8**(2), 50–65 (2004)
51. Raya, M., Hubaux, J.P., Aad, I.: DOMINO: a system to detect greedy behavior in ieee 802.11 hotspots. In: Proceedings of the 2nd International Conference on Mobile Systems, Applications, and Services (MobiSys), Boston, MA (2004)
52. So, J., Vaidya, N.H.: Multi-channel mac for ad hoc networks: handling multi-channel hidden terminals using a single transceiver. In: Proceedings of the 5th ACM Symposium on Mobile Ad Hoc Networking and Computing (MobiHoc), Tokyo, Janpan (2004)
53. Spectrum Bridge, Inc.: http://www.spectrumbridge.com
54. Sweeney, L.: k-anonymity: a model for protecting privacy. Int. J. Uncertain. Fuzz. Knowledge Based Syst. **10**(5), 557–570 (2002)
55. Tzeng, W.: Efficient 1-out-of-n oblivious transfer schemes with universally usable parameters. IEEE Trans. Comput. **53**(2), 232–240 (2004)
56. Varian, H.: Economic mechanism design for computerized agents. In: USENIX Workshop on Electronic Commerce (1995)
57. Vedantham, R., Kakumanu, S., Lakshmanan, S., Sivakumar, R.: Component based channel assignment in single radio, multi-channel ad hoc networks. In: Proceedings of the 12th International Conference on Mobile Computing and Networking (MobiCom), Los Angeles, CA (2006)
58. Vickrey, W.: Counterspeculation, auctions, and competitive sealed tenders. J. Financ. **16**, 8–37 (1961)

59. Wang, W., Li, X., Wang, Y.: Truthful multicast in selfish wireless networks. In: Proceedings of the 10th International Conference on Mobile Computing and Networking (MobiCom), Philadelphia, PA (2004)
60. Wang, W., Eidenbez, S., Wang, Y., Li, X.: OURS: optimal unicast routing systems in non-cooperative wireless networks. In: Proceedings of the 12th International Conference on Mobile Computing and Networking (MobiCom), Los Angeles, CA (2006)
61. Wang, X., Li, Z., Xu, P., Xu, Y., Gao, X., Chen, H.: Spectrum sharing in cognitive radio networks: an auction-based approach. IEEE Trans. Syst. Man Cybern. Part B: Cybern. **40**(3), 587–596 (2010)
62. Wang, S., Xu, P., Xu, X., Tang, S., Li, X., Liu, X.: TODA: truthful online double auction for spectrum allocation in wireless networks. In: Proceedings of the 4th IEEE International Symposium on New Frontiers in Dynamic Spectrum Access Networks (DySPAN), Singapore (2010)
63. Welsh, D.J.A., Powell, M.B.: An upper bound for the chromatic number of a graph and its application to timetabling problems. Comput. J. **10**(1), 85–86 (1967)
64. West, D.B.: Introduction to Graph Theory, 2nd edn. Prentice Hall, Upper Saddle River (1996)
65. Wu, F., Vaidya, N.: SMALL: a strategy-proof mechanism for radio spectrum allocation. In: Proceedings of 30th Annual IEEE Conference on Computer Communications (INFOCOM), Shanghai, China (2011)
66. Wu, F., Vaidya, N.: A strategy-proof radio spectrum auction mechanism in noncooperative wireless networks. IEEE Trans. Mobile Comput. (TMC) **12**(5), 885–894 (2013)
67. Wu, F., Zhong, S., Qiao, C.: Globally optimal channel assignment for non-cooperative wireless networks. In: Proceedings of 27th Annual IEEE Conference on Computer Communications (INFOCOM), Phoenix, AZ (2008)
68. Wu, F., Singh, N., Vaidya, N., Chen, G.: On adaptive-width channel allocation in non-cooperative, multi-radio wireless networks. In: Proceedings of 30th Annual IEEE Conference on Computer Communications (INFOCOM), Shanghai, China (2011)
69. Xu, P., Li, X.: TOFU: Semi-truthful online frequency allocation mechanism for wireless networks. IEEE/ACM Trans. Network. **19**(2), 433–446 (2011)
70. Xu, P., Wang, S., Li, X.: SALSA: strategyproof online spectrum admissions for wireless networks. IEEE Trans. Comput. **59**(12), 1691–1702 (2010)
71. Xu, P., Li, X., Tang, S., Zhao, J.: Efficient and strategyproof spectrum allocations in multichannel wireless networks. IEEE Trans. Comput. **60**(4), 580–593 (2011)
72. Xu, P., Xu, X., Tang, S., Li, X.: Truthful online spectrum allocation and auction in multi-channel wireless networks. In: Proceedings of 30th Annual IEEE Conference on Computer Communications (INFOCOM), Shanghai, China (2011)
73. Yang, D., Fang, X., Xue, G.: Channel allocation in non-cooperative multi-radio multi-channel wireless networks. In: Proceedings of 31st Annual IEEE International Conference on Computer Communications (INFOCOM), Orlando, FL (2012)
74. Yao, A.C.: Protocols for secure computations (extended abstract). In: Proceedings of the 23rd Annual Symposium on Foudations of Computer Science (FOCS), Chicago, IL (1982)
75. Yue, W.: Analytical methods to calculate the performance of a cellular mobile radio communication system with hybrid channel assignment. IEEE Trans. Veh. Tech. **40**(2), 453–460 (1991)
76. Zhang, T., Wu, F., Qiao, C.: SPECIAL: A strategy-proof and efficient multi-channel auction mechanism for wireless networks. In: Proceedings of 32nd IEEE International Conference on Computer Communications (INFOCOM), Turin, Italy (2013)
77. Zheng, H., Peng, C.: Collaboration and fairness in opportunistic spectrum access. In: Proceedings of the 2005 IEEE International Conference on Communications (ICC), Seoul, Korea (2005)
78. Zheng, Z., Wu, F., Chen, G.: SMASHER: a strategy-proof combinatorial auction mechanism for heterogeneous channel redistribution. In: Proceedings of the 14th ACM Symposium on Mobile Ad Hoc Networking and Computing (MobiHoc), Bangalore, India (2013)

79. Zhong, S., Wu, F.: On designing collusion-resistant routing schemes for non-cooperative wireless ad hoc networks. In: Proceedings of the 13th International Conference on Mobile Computing and Networking (MobiCom), Montreal, Canada (2007)

80. Zhong, S., Chen, J., Yang, Y.R.: Sprite, a simple, cheat-proof, credit-based system for mobile ad-hoc networks. In: Proceedings of 22nd Annual IEEE Conference on Computer Communications (INFOCOM), San Francisco, CA (2003)

81. Zhong, S., Li, L.E., Liu, Y.G., Yang, Y.R.: On designing incentive-compatible routing and forwarding protocols in wireless ad-hoc networks: an integrated approach using game theoretical and cryptographic techniques. In: Proceedings of the 11th International Conference on Mobile Computing and Networking (MobiCom), Cologne, Germany (2005)

82. Zhou, X., Zheng, H.: TRUST: a general framework for truthful double spectrum auctions. In: Proceedings of 28th Annual IEEE Conference on Computer Communications (INFOCOM), Rio de Janeiro, Brazil (2009)

83. Zhou, X., Zheng, H.: Breaking bidder collusion in large-scale spectrum auctions. In: Proceedings of the 11th ACM Symposium on Mobile Ad Hoc Networking and Computing (MobiHoc), Chicago, IL (2010)

84. Zhou, X., Gandhi, S., Suri, S., Zheng, H.: eBay in the sky: strategy-proof wireless spectrum auctions. In: Proceedings of the 14th International Conference on Mobile Computing and Networking (MobiCom), San Francisco, CA (2008)

85. Zhu, Y., Li, B., Li, Z.: Core-selecting combinatorial auction design for secondary spectrum markets. In: Proceedings of 32nd IEEE International Conference on Computer Communications (INFOCOM), Turin, Italy (2013)